Lonely 🌐 planet

UNDER
the
STARS
CAMPING
AUSTRALIA &
NEW ZEALAND

THE BEST CAMPSITES, HUTS,
GLAMPING AND BUSH CAMPING

CONTENTS

INTRODUCTION

The call of the wild is perhaps stronger than ever before, with more and more of us longing to escape our daily lives and immerse ourselves in the beauty of the natural world. For some, ideal forays may be weeks in the wilderness, hiking, biking or kayaking from one remote campsite to another. For others, it's a weekend of bush glamping in easy reach of urban life. Everyone's definition of a nature-based escape differs, as does their comfort zone within it. The common denominator is that it feels good. Spending time with Mother Nature is not only a rewarding way to refresh mind, body and soul – science has proven that it is beneficial for our health, too.

Access to nature, and particularly the ability to spend the night within its grasp, is unfortunately not universal. And that was the driving force behind creating this book – we wanted to provide you with an inspiring list of Australia's and New Zealand's most memorable places to drift off to sleep under the stars. Our 200 picks range from the wild and rugged – rustic camps on tropical islands, tents in rolling desert dunes, and walkers' huts on isolated alpine trails – to comfortable safari-style stays and architectural cabins that cling to cliffs and nestle alongside shorelines. Some campgrounds are more traditional with a range of facilities ideal for families, while others require campers to be fully self-sufficient. So whatever your idea of wild is, we have you covered.

For further inspiration, we also delve into some of the best activities that can play a role in your back-to-nature escapes, whether as your primary mode of transport between camps, or simply a way to connect you more deeply to your incredible natural environment between sleeps.

Sarah Reid

HOW TO USE THIS BOOK

The book is organised into nine chapters, one for each state and territory of Australia (the Australian Capital Territory is combined with New South Wales), as well as the North Island and South Island of New Zealand. In each chapter, we highlight the top regions for sleeping under the stars, whether camping, glamping or staying in cabin-style accommodation. We also delve into the rules and regulations of free camping, and discuss safety considerations and where to buy camping supplies locally.

For each of Australia's and New Zealand's under-the-stars sites, our extensive reviews are accompanied with the following practical details: best time to camp, amenities available, contact details and how it is best accessed. Icons (below) also note the cost range of each site.

LEGEND

- Family-friendly
- Environmentally friendly
- Free
- Budget
- Midrange
- Top end

© SAL-SALIS

COME RAIN OR WINE

The rain thunders down on the roof of the car as the windscreen wipers work overtime to grant us but a glimmer of the road ahead with every laboured swipe. Hardly the ideal conditions for camping, we agree, but after two years largely confined to our home in northern New South Wales you-know-when, my outdoors-obsessed husband and I seize the opportunity to cross the border for a camping holiday in Queensland as soon as the opportunity arises.

An ambitious plan to hike a trail and then spend our first night at Girraween National Park (left) saw us hit the road extra early for a three-hour drive west. To our relief, the relentless rain dissipates as we reach the New South Wales Northern Tablelands, and the warm glow of the afternoon sun turns the rural landscape's patchwork of paddocks to gold.

We celebrate the (then-novel) event of crossing the border by dropping into a winery on the way into the national park, which fringes the boutique Granite Belt wine region. Within minutes we're in the wilderness, surrounded by twisted gums and granite boulders, and soothed by the earthy, ever-so-citrussy scent of the Aussie bush. There's a flash of red and blue as a crimson rosella torpedoes through the trees, and our ears prick at the distinctive thump-thump of a startled kangaroo bounding off into the scrub. It's also impossible to ignore the spirituality of this ancient land, used by countless generations of First Nations peoples.

'We should put the tent up before we hike,' says my husband sensibly. An eagle-eyed kookaburra watches us fumble with our new tent like camping rookies (we're not), hopeful of scoring a snack (it doesn't).

We probably should have made a more sensible decision about the hike, too, I rue, as a dodgy cloud dumps its guts just as we hit our stride. Cockatoos screech and wallabies scatter as we make a beeline for the covered veranda of the shuttered information centre.

After half an hour, the downpour still hasn't ceased. We decide to make a dash for our campsite, where we're gutted to discover that in our haste to put up the tent, we hadn't secured the fly properly, and the rain has soaked part of the mattress. Cursing our carelessness, as well as the rain for good measure, we suddenly remember the wine. Mellowed by a glass or two of robust red, we somehow manage to get a barbecue going under an awning, sipping wine as the nostalgic aroma of sizzling sausages and onions fills our nostrils. We're both saturated by the time we chow down on our sausage sandwiches, topped with charred onions and tomato sauce. Yet in this moment I feel more at peace than I have in a long time. That's camping for you. Even when it's a bit of a disaster, it's still a tonic for the soul.

Sarah Reid

AUSTRALIA

From lush tropical islands to rolling red desert dunes, Australia presents infinite opportunities to get off the grid in seriously scenic surrounds.

You don't need to venture far from Australia's urban centres to get wild. With more national parks than any other country (approximately 681 compared to next-in-line China with 225), the country offers a wonderfully diverse collection of protected natural areas to pitch a tent, typically in the company of more native wildlife than fellow campers. Here are a few things to know before you roll out a swag (Australia's favourite tent).

WHEN TO GO

From summer bushfires to alpine blizzards, blistering sun to torrential rain, Australia is a land of extremes. Avoid the harshest climatic conditions by travelling in sync with the seasons. In northern Australia, the two main seasons are the wet (Nov–Apr), which coincides with cyclone and stinger (deadly jellyfish) season, and the dry (May–Oct), which is also cooler and ideal for camping. Central Australia can be dangerously hot in the peak summer months (Dec–Feb), when it's ideal beach weather down south.

With campfires seasonally or permanently banned in many national parks due to bushfire safety measures, it's wise to pick up a portable gas cooker.

SAFETY

Adequate preparation for exploring Australia can save your life. If you're heading off the grid (on foot or by vehicle), a personal locator beacon, plenty of water, extra fuel if driving, snacks and a first aid kit are essential. Always tell someone where you're going, and when you plan to return. If you break down, don't leave your vehicle. Telstra generally has the widest mobile (cell) coverage, but there are lots of black spots.

Deadly wildlife encounters are rare – sadly the ocean and roads claim many more victims – but it's important to know what to do in an emergency; St John Ambulance (www.stjohn.org.au) offers downloadable fact sheets on how to treat everything from snake bites to jellyfish stings. Research local hazards in advance, and always heed safety signage.

FREE CAMPING

Freedom camping (pitching a tent anywhere) is generally banned nationwide, and fines can be steep. But there are plenty of designated areas you can camp for free; the free Campermate app, Hipcamp (www.hipcamp.com) and www.freecampingaustralia.com.au are good resources for finding free and low-cost places to camp in Australia, but always check local council rules before settling in.

CARING FOR COUNTRY

Upholding the leave no trace philosophy while camping in Australia isn't just important for the environment, it also shows respect for the Indigenous Australians who have a deep and ongoing connection to the Country (traditional lands) you're camping on. See a map of Indigenous Australia at www.aiatsis.gov.au.

NATIONAL PARK CAMPSITES

With online booking required for many national park campsites — most of which also lack mobile (cell) coverage — checking ahead is key. National park campsites tend to be very basic; few offer potable water and showers.

NEW ZEALAND

Compact, ever-changing and beautiful beyond measure, New Zealand excels naturally. With an outdoors this good, why would you even want to sleep indoors?

New Zealand is a tightly packed larder of landscapes – there are few countries in the world that can match it, hectare for hectare, for diversity. For campers, this means the possibility of a different environment every night you throw down your tent: a night in the mountains, then coast, lakeside, by a volcano, on an island, or in sight of a glacier. New Zealand was made for camping, so get out there.

WHEN TO GO

Weather moves through fast, and sometimes furiously, on these narrow islands. The warmer months are the best time for camping, especially on the South Island, with winter (Jun-Aug) an unlikely time to want to pitch in the shadow of the snow-smothered Southern Alps. The camping season broadens the further north you head, making for comfortable winter nights out as you head towards Auckland and Northland.

SAFETY

You can all but rule out any animal worries in this country delightfully free of snakes, venomous spiders and anything that looks even vaguely like trouble. Weather and roads are arguably the greatest hazards, so keep a watch on MetService local forecasts (and a vigilant eye on other drivers). If heading remote, carry a personal locator beacon and a first aid kit, and during and after rain be alert to any flash flooding, especially around mountain streams and rivers.

Hato Hone St John (www.stjohn.org.nz) runs first aid courses and has a good online first aid library outlining responses and steps to take in the event of bites, breaks, allergic reactions and other scenarios. Phone coverage is patchy outside of cities and towns, with 2degrees (www.2degrees.nz) providing the widest coverage.

FREE CAMPING

New Zealanders are pretty chilled about life, but not necessarily about where you camp. There are around 500 designated freedom camping sites around the country and it's best to stick to them if you're planning to camp for free. Many freedom camping sites, especially around settlements, are exclusively for self-contained vehicles, and they are routinely monitored, with fines applying. The camping map at https://camping-nz.rankers.co.nz is a good starting point to locate free sites, as is the free CamperMate app.

NATIONAL PARK CAMPSITES

The Department of Conservation (DOC) manages more than 200 campsites in parks and reserves around New Zealand. DOC campsites are divided into five categories, from basic (free) campsites through to serviced campsites with flush toilets and tap water. A Campsite Pass can be purchased for 30 days or 365 days, and can be used for up to seven nights in a 30-day period. Many campsites need to be booked ahead, which can be done online at www.doc.govt.nz.

Maps showing DOC campsites by region or island can be downloaded from the website.

NEW SOUTH WALES & ACT

You don't need to venture too far from the urban grind of Sydney or Canberra to soak up sparkly skies and nature vibes.

Best time: year-round (North Coast, country NSW), Sep–Apr (South Coast, ACT, alpine), May–Nov (outback)
Best national parks for camping: Kosciuszko National Park, Mimosa Rocks National Park, Murramarang National Park
Best camping trails: Australian Alps Track, Gidjuum Gulganyi Walk, Green Gully Track
National parks pass required: Yes
Useful contacts: www.visitnsw.com; www.nationalparks.nsw.gov.au; https://visitcanberra.com.au; www.parks.act.gov.au

The Sydney Harbour Bridge and Opera House are defining images of New South Wales. But you only need to gaze across the glittering harbour backdropping Sydney's architectural icons to get a taste of the state's staggering natural beauty. From the otherworldly geological formations of Mungo National Park in the depths of the outback to the teeming Gondwana Rainforests stretching from the Queensland border to the edge of the Hunter Valley, there is an endless supply of opportunities to get back to nature. Between wild walks and wildlife spotting, there's culture to uncover, too, with rock art tucked in ancient gorges and middens spilling out of sand dunes among the physical reminders of the ongoing connection the state's Traditional Custodians share with this storied Country.

National parks cover more than 10 per cent of New South Wales and almost half of the Australian Capital Territory, with hundreds of campgrounds between them. And you barely need to start the car in either of their capital cities to find a tiny cabin or glamping stay that is designed to connect you with nature in perfect comfort.

FREE CAMPING

Scenic free camping spots are tricky to find in Australia's most populated state, where many councils have banned it. In the Kangaroo Valley south of Sydney, the riverside Bendeela Recreation Area is also popular with wombats. Some national parks have 'free' camping areas, but a $6 booking fee applies.

SUPPLIES

The big brands are well represented in larger cities including Sydney, Canberra, Newcastle, Wollongong and Coffs Harbour. Try the independent Byron Bay Camping & Disposals (www.byron-camping.com.au) on the far North Coast.

SAFETY

With more than 50 lives lost to coastal drownings in New South Wales annually, it pays to be 'beachsafe' (https://beachsafe.org.au). Monitor bushfires and fire bans via the Hazards Near Me app (www.nsw.gov.au/emergency/hazards-near-me-app).

BEST REGIONS

North Coast

Gorgeous beaches, lush rainforests, warm (enough) weather year-round. What more

Eco huts and hot tubs at Kimo Estate
(left); a watchful kookaburra (below)

do you need to entice you to pitch a tent? Alright, we'll throw in some good waves and great regional restaurants, too.

South Coast

Slip into something warm (sometimes even in summer) and camp along the rugged coastlines south of Sydney, which only get wilder as you head further south. Swim in hidden bays, commune with wildlife, and find inner calm beside crackling fires.

Country New South Wales

Roamed by livestock and 'roos, and dotted with national parks, wineries and cultural sites, the beating country heart of New South Wales is full of friendly faces and big-sky vistas.

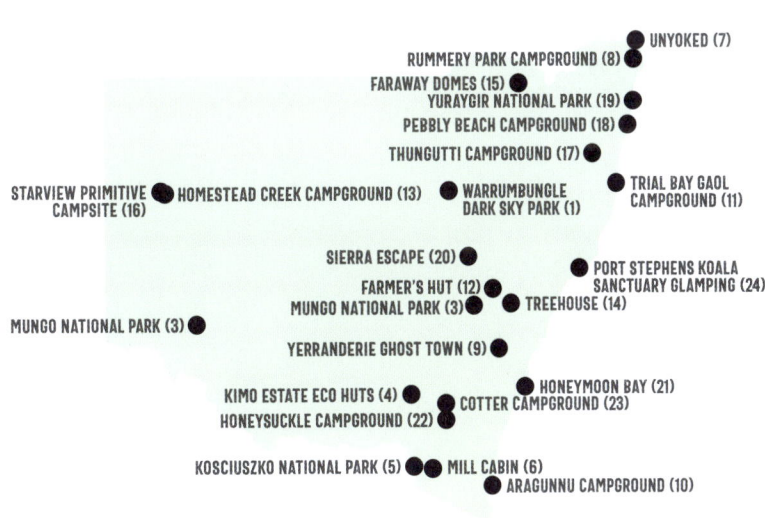

UNYOKED (7)
RUMMERY PARK CAMPGROUND (8)
FARAWAY DOMES (15)
YURAYGIR NATIONAL PARK (19)
PEBBLY BEACH CAMPGROUND (18)
THUNGUTTI CAMPGROUND (17)
STARVIEW PRIMITIVE CAMPSITE (16)
HOMESTEAD CREEK CAMPGROUND (13)
WARRUMBUNGLE DARK SKY PARK (1)
TRIAL BAY GAOL CAMPGROUND (11)
SIERRA ESCAPE (20)
PORT STEPHENS KOALA SANCTUARY GLAMPING (24)
FARMER'S HUT (12)
MUNGO NATIONAL PARK (3)
TREEHOUSE (14)
MUNGO NATIONAL PARK (3)
YERRANDERIE GHOST TOWN (9)
HONEYMOON BAY (21)
KIMO ESTATE ECO HUTS (4)
COTTER CAMPGROUND (23)
HONEYSUCKLE CAMPGROUND (22)
KOSCIUSZKO NATIONAL PARK (5)
MILL CABIN (6)
ARAGUNNU CAMPGROUND (10)

WARRUMBUNGLE DARK SKY PARK
CENTRAL WEST

If there's one stargazing spot in New South Wales to trump all others, it's Warrumbungle National Park near Coonabarabran, where a combination of dark skies, low humidity and high altitude supported its certification as Australia's first Dark Sky Park in 2016. Get a closer look through the lenses of two eight-inch Dobsonian telescopes and binoculars on a ranger-led Explore the Dark Sky tour, typically held during school holidays. Though on a cloudless night you don't really need any help to admire the Milky Way in mesmerising detail – especially between February and October, when its galactic centre is on full display in the Southern Hemisphere.

Choose from a dozen campgrounds in the 2WD-accessible national park, with the largest, Camp Blackman, popular for its hot showers, barbecues and resident kangaroos. It's also close to the visitor centre, where astronomy tours begin. Feel the power of the landscape as you explore the national park during the day, where a dramatic mix of volcanic spires and domes, plateaus, forested ridges and tall volcanic dykes (including the distinctive Breadknife) provide a spectacular backdrop for astrophotography. With each campground providing unique angles, it's worth sampling a few to maximise your portfolio. Rug up – it can get chilly out here.

01

🧭 THE PITCH

Camp under the starriest of skies in this ancient volcanic landscape northwest of Sydney, famously home to the nation's first internationally accredited Dark Sky Park.

When: Feb–Oct
Amenities: most campgrounds have toilets and picnic tables
Best accessed: by car
Cost: from $6
Contact: www. nationalparks.nsw.gov.au; https://darksky.org

BUBBLETENT AUSTRALIA
CAPERTEE VALLEY

Marvel at the Milky Way from the comfort of your queen bed in an inflatable 'bubbletent' overlooking the immense Capertee Valley, under three hours' drive from Sydney. Each named after a star sign, these three transparent domes nestling on a 400-hectare working farm offer the same facilities including an ensuite, outdoor wood-fired hot tub, outdoor gas-powered camp kitchen and a telescope for peering deep into the night sky. Virgo also has ducted air-con to take the edge off summer nights.

Spend a quiet day relaxing in the hammock with your book, canoodle on the love swing,

or take a drive to the wineries of nearby Mudgee. Savour a luxuriously long soak in your tub as the sun goes down, then whip up an alfresco meal. Play around with the Luminous app on the in-bubble iPad and see what Earth

looks like from Saturn's POV. Or master the art of photographing a close-up of the moon through your personal telescope. Simply bring food, drinking water and your favourite person, and experience a new way to spend a night under the stars, warmed by the electric blanket. The solar-powered bubbles don't have power points, but USB charging is available.

 **THE PITCH**

Gaze out across the world's widest canyon by day, then up to the stars at night, from this trio of off-grid domes on the edge of the Blue Mountains World Heritage Area.

When: year-round
Amenities: private bathroom, kitchen, firepit
Best accessed: by car
Cost: $1430 for 2pp for 2 nights
Contact: https://bubbletentaustralia.com

03

MUNGO NATIONAL PARK

WILLANDRA LAKES REGION

One of the world's most culturally significant landscapes, Mungo National Park, lies 1000km (621 miles) west of Sydney. The world's oldest ritual burials were performed in this place some 42,000 years ago. Along with 20,000-year-old fossilised human footprints, they tell an incredible story of the long history of Australia's First Nations.

There are two spots to camp in this ancient landscape. Just a short walk from the visitor centre (which has hot showers), the Main Campground makes a great base for exploration. For more solitude, pitch up among grazing kangaroos at Belah Campground, set roughly halfway along the 70km (43.5-mile) Mungo self-guided drive. Just outside the park, Mungo Lodge has a restaurant (and rooms).

Seize the opportunity to visit the park's famous Walls of China on an Aboriginal ranger-guided walk. Learn how these compacted sand and clay formations are rich time capsules, revealing historical treasures including megafauna skeletons and remnants of ancient kitchen fires. And learn about fascinating finds such as the ancient footprints on a clay pan, which scientists believe belonged to hunters. A 4WD is an asset in May, the park's wettest month.

THE PITCH

You may as well be camping on the moon in this remote outback World Heritage Site, where surreal desert landscapes shaped by the elements intertwine with Aboriginal heritage.

Best time: Apr–Nov
Amenities: toilets, picnic tables, plus barbecues at the main campground
Best accessed: by car
Cost: from $12
Contact: www.nationalparks. nsw.gov.au

© RUGLIG | SHUTTERSTOCK

KIMO ESTATE ECO HUTS
GUNDAGAI

Get a taste of minimalist luxury in one of Kimo Estate's architect-designed, adults-only 'eco huts', each privately positioned on a 2833-hectare farm just outside the small country town of Gundagai, southwest of Sydney. Enjoy a lie-in as the sun streams in through the large windows of your off-grid hut, perched on its own hill overlooking the Murrumbidgee River flats, and cap off the day with a leisurely soak in the wood-fired outdoor hot tub. Prepare a feast in the alfresco kitchen to enjoy on your private deck, and fire up the wood-fired heater inside on chilly nights. The solar-powered huts don't come furnished with a fridge, but there's an esky to keep your food cold, including provisions kindly provided by your hosts for a hearty local-produce-led breakfast.

Head out for a walk on the property to meet the resident sheep, or enjoy the solitude of your own off-grid space, constructed with locally sourced timber that helps to create a warm and inviting ambience. In the summertime, the hot tub transforms into a refreshing plunge pool.

Another two cottages and a converted shearers' quarters on the same property offer similarly atmospheric accommodation.

04

© MATT BEAVER PHOTOGRAPHY | KIMO ESTATE

 **THE PITCH**

Unplug in style at one of these three designer A-frame huts in the rolling hills of country New South Wales, each with an alfresco tub.

When: year-round
Amenities: drinking water, private bathroom, electricity, wi-fi
Best accessed: by car
Cost: $600
Contact: www.kimoestate.com

KOSCIUSZKO NATIONAL PARK
SNOWY MOUNTAINS

Stretching across the vast alpine landscape of southern New South Wales, Kosciuszko National Park is the state's largest protected area. As the snow blanketing its peaks and valleys gives way to wildflower-lined trails and trout-filled rivers in the spring, opportunities for adventure open up well beyond its four ski resorts.

More than 40 campgrounds (and an additional 10 national parks-run lodges) dot the park. Selecting a spot for a night under the stars depends largely on how you want to spend your time. Shaded by snow gums alongside the Thredbo River, the popular Thredbo Diggings Campground is a great base for hiking, biking and fishing. Unsaddle on a multi-day riding adventure on the National Trail at Geehi Horse Camp, or pack up your 4WD to enjoy the serenity of the solitary campsite at Buddong Falls Campground.

With the handful of historic huts that remain in the park only to be used for temporary day use and emergencies, the two-bedroom Daffodil Cottage is arguably the most atmospheric place to stay in the park with an actual roof over your head. Once the overseer's residence for the historic Currango Homestead, it offers a glimpse into the pastoral history of this remote alpine landscape.

THE PITCH

Home to Australia's highest peak, the rugged alpine wilderness of the New South Wales high country brings you closer to the stars than anywhere else on the continent.

When: Oct—May
Amenities: varies
Best accessed: by car
Cost: from $6
Contact: www.national parks.nsw.gov.au

05

© TETRA IMAGES | GETTY IMAGES

MILL CABIN
SNOWY MOUNTAINS

06

Step back to Australia's pioneer days – albeit in a very modern way – at Mill Cabin, nestling on a private bush property near Jindabyne in the New South Wales Snowy Mountains. With its pitched-roof design and beautiful granite rock walls reminiscent of a traditional stockman's hut, this off-grid one-bedroom cabin blends seamlessly into the woodlands of the sub-alpine landscape. Inside, indulge in the little luxuries of a lovingly curated modern home: designer eco-friendly toiletries, plush Australian-made bedding, a well-appointed kitchen and a bookshelf stuffed with holiday reads and board games to compensate for the absence of TV. From the custom ceramic dinnerware to the timber joinery, almost all of the furnishings were conceived specifically for the cabin, with raw and natural materials prioritised to create a deliberate connection with the surrounding landscape.

Cosy up inside as snow blankets the landscape at the peak of winter, or strike out to explore the ski fields of Perisher and Thredbo. When the snow thaws, there are plenty of hiking and biking trails to discover. A Japanese-inspired outdoor bathtub offers a restorative soak as you gaze across the valley towards Mt Perisher, scanning the rolling hills in between for kangaroos, wallaroos and deer.

🍃 THE PITCH

Embrace a simpler way of life on a stay at this private, modern bush cabin inspired by the historic stockman huts that once covered the New South Wales High Country.

Best time: year-round
Amenities: drinking water, kitchen, private bathroom, electricity, wi-fi
Best accessed: by car
Cost: $900 for 2 nights
Contact:
www.millcabin.com.au

07

UNYOKED
VARIOUS LOCATIONS

Nothing makes your cortisol levels drop quite like connecting with nature. That was the thinking behind the creation of Unyoked, a collection of more than 100 off-grid tiny cabin stays located in secluded corners of private properties across Australia, New Zealand and also the UK. With a queen-sized bed framed by huge windows and a generous front deck, each cabin has the same compact design and eco-friendly footprint. Some cabins, like Pana and Ashi – surrounded by towering eucalypts in the New South Wales Northern Rivers – come with extra amenities including a wood-fired outdoor

bath. Like many other Unyoked stays, these two cabins also have a firepit, perfect for roasting marshmallows as wallabies and bandicoots snuffle around in the shadows. Locally sourced provisions are another welcome touch in every cabin.

There's a bit of mystery to an Unyoked stay, with the exact location not revealed until

shortly before your visit. When booking, you can choose your perfect cabin based on photos, the general location, strength of mobile/cell reception, and its calm, creativity and clarity levels, each rated from one to three. When you arrive, use the barrows to transport your gear to your wild stay.

🌿 THE PITCH

Pen a bestseller – or simply read one – on a stay at one of these stylish tiny cabins dotted around New South Wales and beyond, designed to facilitate connection with nature and inspire creativity.

When: year-round
Amenities: drinking water, private bathroom, toilet, fridge, kitchenette, electricity
Best accessed: by car, then on foot
Cost: from $566 for 2 nights
Contact: www.unyoked.co

© TRENTANDJESSIE | UNYOKED

RUMMERY PARK CAMPGROUND
WHIAN WHIAN STATE CONSERVATION AREA, NORTHERN RIVERS

Nearly 50 years ago, the ancient rainforests of northeastern New South Wales hosted one of Australia's most fiercely fought conservation battles. In a loss for the sawmills, Nightcap National Park was declared in 1983, and World Heritage listed in 1989. In 2003, the neighbouring Whian Whian State Conservation Area was created to enhance its protection.

Plunging 100m (328ft) off the rim of ancient shield volcano is Nightcap's showpiece Minyon Falls. It's only a 45-minute drive from the popular beach town of Byron Bay, but to fully immerse in the magic of this rainforest, you need to stay overnight. From the Minyon Falls day-use area, it's only 2km (1.2-miles) by road to the Rummery Park Campground, just inside the Whian Whian

State Conservation Area, with 18 grassy sites for tents, trailers and caravans. Spend a leisurely day hiking to the base of Minyon Falls, or hit the Whian Whian mountain biking trails. Scan the surrounding gum trees for koalas as you fire up a barbecue feast, but don't take your eyes off the hotplate because the local kookaburras are always watching.

Scheduled to open in 2024, the four-day Gidjuum Gulganyi Walk will also thread through this beautiful Bundjalung Country landscape.

08

© SEETHEWORLD PHOTOGRAPHY | SHUTTERSTOCK

🟢 THE PITCH

Say a quiet thanks to the conservationists of the 1970s who helped protect this glorious tract of rainforest, giving you a new appreciation for the colour green in the area's only campground.

Best time: year-round
Amenities: toilets, barbecues, picnic tables
Best accessed: by car
Cost: from $24.60
Contact: www.national parks.nsw.gov.au

YERRANDERIE GHOST TOWN
YERRANDERIE REGIONAL PARK

Travel back to the early days of Federation in the former silver mining town of Yerranderie, home to more than 2000 people at its peak in 1911. Now occupied by more kangaroos than human residents, it's one of Australia's best-preserved colonial ghost towns. You can spend a night here, with three of Yerranderie's rustic buildings and two campgrounds providing accommodation just steps from the heart of the township. With just one bedroom, the tin-roofed Bank Room, still displaying its original signage, offers a particularly intimate experience.

Get a sense of daily life here a century ago as you stroll main street, now littered with mining memorabilia. Take short walks in the surrounding bushland, go 4WD touring along the Oberon Colong Historic Stock Route, and peer down into the abandoned Silver Peak Mine from the bridge installed across its shaft.

Getting here involves bumping along 76km (47 miles) of dirt road, which only highlights the sense of discovering this settlement that time forgot. Following the decline of the local mining industry, Yerranderie's death knell came in the 1950s, when the establishment of the Warragamba Dam and Lake Burragorang cut off its direct access to Sydney to the northeast.

09

● THE PITCH

An abandoned silver mining town sets the scene for an eerie night camping in Yerranderie Regional Park or bedding down in one of its restored historic buildings.

When: year-round
Amenities: drinking water, toilets, showers, barbecues, picnic tables; toilets only at Government Town Campground
Best accessed: by car
Cost: camping from $6
Contact: www.national parks.nsw.gov.au

ARAGUNNU CAMPGROUND
MIMOSA ROCKS NATIONAL PARK

Hugging the sparkling Sapphire Coast north of Tathra, Mimosa Rocks National Park is a real coastal camping gem, with four campgrounds offering a scenic immersion in the littoral rainforest of Yuin Country. But the Aragunnu Campground is particularly special.

The 31 campsites in this 2WD-accessible ground are spread across four distinct camping areas lining the 1km (0.6-mile) Mimosa Rocks walking track, which straddles the park's largest Aboriginal midden. The sweetest spot is at the northern end, which offers easy access to the dramatic rocky headlands that claimed the paddle steamer *Mimosa* back

in 1863. There's good snorkelling off the rocky beaches here, while the park's headlands are great vantage points for whale-watching in winter. But you don't need to step outside your campsite for superb wildlife-viewing, with swamp wallabies commonly seen grazing in the grassy camping areas. After dark, you might even spy a wombat or a long-nosed potoroo. The park also provides a refuge for koalas and ringtail possums, and while you may not see yellow-bellied gliders flying between trees, you might be able to hear their distinctive cackling that cuts through the peaceful sounds of the sea.

● THE PITCH

A cocktail of natural splendour is served up at this rustic seaside campground, where wallabies typically outnumber fellow campers, and sugar gliders drift through the trees.

When: Sep–Apr
Amenities: toilets, barbecues
Best accessed: by car
Cost: $24.60
Contact: www.nationalparks.nsw.gov.au

10

TRIAL BAY GAOL CAMPGROUND
ARAKOON NATIONAL PARK

It's not every day you get to camp beside a historic prison. Touring the eerie, convict-built Trial Bay Gaol is just one of the highlights of camping in Arakoon National Park, which wraps around a scenic headland near the town of South West Rocks. Join a ranger-guided tour at twilight and learn about the darker side of the prison's history; from bushrangers to Germans interred during WWI, these sandstone walls have seen it all.

With no less than 97 sites, this campground isn't exactly intimate, but it's got a gorgeous outlook across Trial Bay, with room for all types of camping, and powered sites available. The extra amenities make it a popular destination for families, with the local eastern grey kangaroos providing plenty of entertainment for all.

Follow the trail looping the headland for a refreshing dip at pretty Little Bay. South West Rocks is also one of the state's most famous scuba diving destinations, considered one of the best places in the world to observe critically endangered grey nurse sharks year-round.

If you prefer a more secluded sleep under the stars, Hat Head National Park's basic Smoky Cape Campground is just a 10-minute drive to the south.

● THE PITCH

A state heritage-listed prison with a horrifying history looms above this scenic coastal campground with all the trimmings, plus kangaroos galore.

When: year-round
Amenities: drinking water, toilets, showers, barbecues, electricity, picnic tables
Best accessed: by car
Cost: $35
Contact: www.national parks.nsw.gov.au

© TARAS VYSHNYA | SHUTTERSTOCK

12

FARMER'S HUT
BATHURST

© FARMER'S HUT

Experience a different kind of farm stay at Wilga Station, just outside Bathurst in the New South Wales Central Tablelands, where a secluded hut with a roof of grass sits in perfect harmony with the lush green hill it perches atop on this 105-hectare working sheep station. Inside the open-plan Farmer's Hut you'll find a king-size bed, wood fireplace, bathroom, and a kitchenette for prepping a gourmet breakfast with the local produce provided in a complimentary hamper.

Follow the winding Wilga Sheep Trail to willow-lined Wilga Creek to spend a relaxing day lazing on its banks, soaking up the sunshine. Or breathe in the crisp, lanolin-scented country air on a walk around the property. Savour a bottle of premium local wine (included in your hamper) beside the outdoor firepit as the sinking sun sets the rolling hills aglow,

then snuggle up beside the indoor fireplace as darkness falls.

Treat yourself to a lie-in as the morning mist swirls around the undulating landscape, which could be dusted with snow in mid-winter. Aside from the owners, the only other people you're likely to cross paths with on your stay are guests of the farm's Shearer's Hall accommodation.

THE PITCH

This secluded couples retreat on a historic sheep farm puts a unique luxury spin on a classic Australian farm stay, with local goodies — and plenty of wide open spaces — included.

When: year-round
Amenities: drinking water, kitchenette, private bathroom, electricity, firepit
Best accessed: by car
Cost: $900 for 2 nights
Contact: https://wilgastation.com.au

Surfing

The birthplace of Australian surfing, New South Wales serves up waves for complete novices through to seasoned shredders all the way along its glittering Pacific coast. Get frothing.

Hawaii's Olympic swimming champion Duke Kahanamoku is often credited with introducing surfing to Australia in the summer of 1914-15 at Freshwater Beach in Sydney, yet photos show Sydneysiders – male and female – riding waves years earlier. What can be agreed on is that Australia's surfing history kicked off in the New South Wales capital, and from Palm Beach at its northern tip to Cronulla in the south, Sydney's waves are still smokin'. But with nearly 900 recorded beaches fringing the state's Pacific coastline, there are so many more breaks to discover. If you're just starting out, you'll find a surf school in just about every beach town.

Winter swells typically deliver the biggest waves, but Huey (the God of surfing) is known to bless surfers with plenty of sweet summertime sessions. If a northeasterly wind is blowing, check the sand for bluebottle jellyfish before paddling out into a potential stinger-fest. And be aware that river mouths can be sharky, especially after rainfall.

THE SEARCH

Surfing is all about the search, and with one of the world's most surfable coasts, New South Wales was made for hunting the perfect wave. While few of the state's secret spots remain truly secret these days, it's still possible to score uncrowded waves on less

populated sections of the coast, especially if there's an option to camp right in front of the break. So pack a board with your tent, and hit the road.

SURFERS' CODE

The biggest breach of the surfers' code is dropping in (not giving way). Always look before you take off to ensure you don't cut off a fellow surfer. The other big one is not wearing a leg rope. Following a near-fatal accident in Byron Bay in 2023, you can now cop a fine for surfing at its beaches without one. Sticking to waves that match your ability will also help to keep you and fellow beachgoers safe.

Assessing the surf at Manly and Byron Bay (above) and Noosa Heads (left); carving across a Byron Bay wave (inset)

FIVE TO TRY

Byron Bay
Right-handers at The Pass, surfing with dolphins at Wategos: it's clear why Byron became world-famous.

Angourie
This right-hand headland break just south of Yamba produces fast, tubing waves in ideal conditions.

Crescent Head
Another popular right-hander, with sleepy North Coast beach-town vibes to match.

Manly
Paddle out where it all began at this fun beach break at the gateway to Sydney's Northern Beaches.

Gerroa
The long rolling waves of Seven Mile Beach, just south of Kiama on the South Coast, are ideal for learning.

HOMESTEAD CREEK CAMPGROUND
MUTAWINTJI NATIONAL PARK, OUTBACK

The arid plains, rocky gorges and red-dirt tracks of Mutawintji National Park, on the traditional lands of the Malyankapa and Pandjikali peoples, have been an important meeting place for local cultures for tens of thousands of years, with initiations, rainmaking and other ceremonies traditionally performed here.

Gain a deeper understanding of the park's cultural significance with a stay at its only campground, Homestead Creek. This large, shady and well-equipped campground is suitable for tents through to motorhomes. Follow the short, wheelchair-accessible pathway through the splendour of Mutawintji's gorges to the rocky overhang Thaaklatjika (Wright's Cave) where you'll find paintings, stencils and engravings that depict pre- and post-colonial Aboriginal history. Be awed yet again by rock art on a tour with Mutawintji Heritage Tours that takes you into a restricted area of the Mutawintji Heritage Site. To connect further with outback Aboriginal cultures, visit during the annual Mutawintji Cultural Festival in August.

● THE PITCH

This dusty creekside campground, a 90-minute drive from Broken Hill, plays host to one of the state's premier outback Indigenous festivals each winter.

When: May–Nov
Amenities: toilets, showers, barbecues, picnic tables
Best accessed: by car
Cost: $12.30
Contact: Contact: www.nationalparks.nsw.gov.au

© GH PRODUCTIONS | SHUTTERSTOCK

14

TREEHOUSE
BLUE MOUNTAINS

Breathe in the fresh, eucalyptus-scented air as you gaze out across the treetops of the vast Blue Mountains, west of Sydney, from your very own treehouse. Nestling on 243 hectares of private wilderness sandwiched between the Blue Mountains National Park and Wollemi National Park near Bilpin, this secluded, one-of-a-kind cabin puts your queen-size bed at eye-level with the bevy of birds that call these secluded areas home. There's an in-ground spa bath positioned next to the floor-to-ceiling window, and with its twisting natural log beams, corrugated iron roof and 'keep out, no grownups' sign on the front door, the rustic, open-plan cabin inspires a sense of child-like wonder. This quirky retreat is strictly for adults only though.

Venture out by day to trace national-park hiking trails, or at night to marvel at the glow worms clinging to the cliffs surrounding nearby Bulcamatta Falls. Or simply get the wood-burning fire going, sink into the spa bath, and savour the views.

The Treehouse forms part of a charming collection of nature-based cabin stays in the area, which also includes the Enchanted Cave, constructed on a rocky outcrop with its own set of beautiful bushland views.

THE PITCH

Channel your inner child on a whimsical stay in a treehouse-style cabin high in the eucalypt forests of the World Heritage-listed Blue Mountains.

When: year-round
Amenities: drinking water, private bathroom, kitchenette
Best accessed: by car
Cost: $2280 for 2pp for 2 nights
Contact: www.lovecabins.com.au

FARAWAY DOMES
NORTHWESTERN NEW SOUTH WALES

Revel in the luxury of solitude at Faraway Domes, a twin set of light-filled, geodesic domes privately situated on a 3642-hectare property around two hours west of Glen Innes. You can revel in actual luxury, too – each thoughtfully furnished, elevated dome has its own private plunge pool and outdoor bathtub as well as an ensuite. Enjoy a sleep-in before taking a leisurely bushwalk, or spend a day relaxing on your timber deck before cooking up a feast in the open-plan kitchen. Kwiambal National Park's scenic Macintyre Falls is among a handful of attractions located within an hour's drive, but staying put and enjoying the simple pleasure of unplugging in the countryside is difficult to resist.

Made up with Egyptian cotton linen, the four-posted king bed offers a peaceful slumber after the sun finally sinks in ribbons of pastel pink. But it's worth staying up to stargaze from your deck, or in front of the cracking fire pit just outside your dome. With no towns of size within miles, it's a light show you won't want to miss. Warm night? There's air-con if you need it, with the domes engineered both to withstand the harsh landscape and maximise energy efficiency.

THE PITCH

Feel a million miles away from civilisation at this pair of adults-only, off-grid glamping domes, perched on a remote ridgeline overlooking the rugged countryside just south of the Queensland border.

When: year-round
Amenities: drinking water, private bathroom, kitchen
Best accessed: by car
Cost: $1050 for 2pp for 2 nights
Contact: www.farawaydomes.com

15

© FARAWAY DOMES

STARVIEW PRIMITIVE CAMPSITE
LIVING DESERT STATE PARK, BROKEN HILL

Tucked in the Barrier Ranges just 12km (7.5 miles) from the legendary outback city of Broken Hill, the Living Desert State Park is best known for its Sunset Sculptures carved from 53 tonnes of sandstone by local artists in 1993. The predator-proof John Simons Flora and Fauna Sanctuary is another highlight, with two short walking trails designed to connect visitors with the region's rich natural and cultural heritage. Less well known is the desert park's excellent council-run campsite offering a more tranquil alternative to Broken Hill's caravan parks.

Pull into one of the 15 unpowered gravel van sites or pitch a tent in the wood-chipped walk-in tent camping area. Each section has its own barbecue area, with an amenities block and 'star-view seating' in between. You'll be provided with an entry code for the gate when you book online; have a card handy to pay the one-off $6 entry fee to enter the desert park. As you can guess from the name, wandering the trail connecting the Sunset Sculptures is a lovely way to spend the waning daylight hours. There are no fires allowed in the campground, but with a sky this sparkly to gaze at, you may not even notice.

16

© HILKE MAUNDER | ALAMY STOCK PHOTO

🗨 THE PITCH

Leaning into the isolation of the outback? On a road trip to Broken Hill, this well-maintained desert campground offers a peaceful retreat from the city.

When: Mar–Nov
Amenities: untreated water, toilets, showers, barbecues, picnic tables
Best accessed: by car
Cost: $12
Contact: www.brokenhill. nsw.gov.au

17

THUNGUTTI CAMPGROUND
NEW ENGLAND NATIONAL PARK, NORTHERN TABLELANDS

Wake up at Thungutti Campground misty rainforest wonderland, where the Gondwana Rainforests of New England National Park take on a particularly ethereal quality some 1300m (4265ft) above sea level. Here, vivid lichens creep across cool rocks, old man's beard drips from gum trees, and after chilly nights the landscape sparkles with a fresh coat of frost.

Follow the lush Tea Tree Falls Walking Track or take a short drive to Point Lookout for energising views across the ancient rainforest from the edge of the Great Escarpment – the sunrises here are incredible. Then take your pick from a handful of trails leading deeper into the wilderness. Listen for the distinctive mimicking call of the superb lyrebird on the Lyrebird Walking Track (5.1km/3.2-mile loop) or chase waterfalls framed by ferns on the Cascades Walking Track (5.7km/3.5-mile loop).

Even the 15-minute Point Lookout Walking Track offers an intriguing taste of this enchanting ecosystem, as grey goshawks ride the skies above. The park is also home to the mostly nocturnal spotted quoll, giving campers the best shot at spying one of these elusive creatures. Seeking more comfort? There's cabin accommodation inside the park.

⊖ THE PITCH

High up in the Northern Tablelands, the only formal campground in New England National Park promises misty mornings, hiking trails, and wildlife-spotting opportunities galore.

When: Sep–Apr
Amenities: toilets, barbecues, picnic tables
Best accessed: by car
Cost: $12.30
Contact: www.nationalparks.nsw.gov.au

© TYRRANNOID | SHUTTERSTOCK

PEBBLY BEACH CAMPGROUND
MURRAMARANG NATIONAL PARK, SOUTH COAST

Spotted gums grow right down to the azure ocean in Murramarang National Park near Batemans Bay on the Eurobodalla coast, offering plenty of shade and a stunning backdrop for a camping adventure. Spanning 44km (27 miles) of dramatic shoreline, the park is chock-full of native wildlife, with eastern grey kangaroos among the most commonly spotted locals. They're often seen grazing and lazing at the picnic area above pretty Pebbly Beach – right beside one of the park's most scenic campgrounds. What was that flash of orange and green? Probably a king parrot, flapping between the flowering trees.

With more than a dozen walking trails threading through the coastal rainforest, keen hikers will love this camping trip. But beach-lovers aren't forgotten, with many of the park's trails weaving past, protected bays only accessible on foot. Among the top walks is the recently launched 34km (21-mile) Murramarang South Coast Walk. Designed to be hiked over three days, walkers can stay in campgrounds or cabins at Depot Beach and South Durras. This doesn't necessarily mean Pebbly Beach Campground will be quieter, but even when its 23 campsites are full, it still provides 2WD-accessible natural healing.

⊖ THE PITCH

Bouncing back from Australia's 2019-20 bushfire season, Murramarang National Park invites you to camp among kangaroos just a short stroll from some of the state's loveliest beaches.

When: Sep–Apr
Amenities: drinking water, toilets, showers, barbecues, picnic tables
Best accessed: by car
Cost: $24.60
Contact: www. nationalparks.nsw.gov.au

18

YURAYGIR NATIONAL PARK
MID NORTH COAST

Stretching north of the regional centre of Coffs Harbour to the peaceful fishing town of Yamba, Yuraygir National Park protects a stellar sweep of the Mid North Coast. With only a handful of teeny coastal towns interrupting this 65km (40 miles) stretch of rocky headlands, striking red cliffs, isolated beaches and quiet lakes set against a backdrop of coastal forests, heaths and bird-filled wetlands, it serves up a whole lotta nature just a short detour off the Pacific Highway.

Eight campgrounds are dotted along the park's shores, each with its own unique setting. For riverside bliss, head to Sandon River, where the campground perches on a sandy peninsula. Want waves? Try Illaroo, where the swell curls around a headland, and the rock pools of nearby Minne Water Bay provide an idyllic setting for a swim. Need some alone-time? The walk-in Rocky Point Campground has you covered.

Naturally, there are plenty of trails to explore, not least the epic Yuraygir Coastal Walk, which follows the entire length of the park. Visit in spring to see the landscape pop with colourful wildflowers, or drop by in winter to watch whales bounce along the coast as white-bellied sea eagles glide overhead.

19

THE PITCH

Choose your own coastal bush camping adventure in this large national park where walks, wildlife, waves and whales are just a few highlights encouraging visitors to linger.

When: year-round
Amenities: toilets, plus drinking water, barbecues and picnic tables at some sites
Best accessed: by car
Cost: from $24.60
Contact: www. nationalparks.nsw.gov.au

© JAKUB MACULEWICZ | SHUTTERSTOCK

SIERRA ESCAPE
MUDGEE

Rouse yourself to admire the rolling hills of the Mudgee region, northwest of Sydney, bathed in golden light at Sierra Escape, a collection of five adults-only glamping tents and one tiny house on a private property just 20 minutes from town. With a generous 54 sq m (528 sq ft) footprint, the four premium tents, each furnished with a king-size bed, give you all the space you need to unwind in the countryside. While the Carinya tent is more compact, this newer addition to the Sierra family features the same luxe amenities including an indoor fireplace and a standalone bathtub on the outdoor deck – with another tub indoors. Meticulously crafted from a shipping container, Elouera, the property's first tiny house, follows suit, with even larger windows drawing the eye across the rippling landscape.

Opt to add a local-produce platter, in-room massage or even a private yoga class. Or venture out to sip your way around Mudgee's cool-climate wineries and dine at its award-winning restaurants before cosying up around your firepit under the big skies of the New South Wales countryside.

THE PITCH

When you want to be close to nature as well as wineries, this string of glamping stays gives you both, along with a little luxury.

When: year-round
Amenities: drinking water, private bathroom, kitchen, firepit, electricity, wi-fi, swimming pool
Best accessed: by car
Cost: from $1280 for 2 nights
Contact: www.sierraescape.com.au

20

© SIERRA ESCAPE

21

HONEYMOON BAY

JERVIS BAY, SHOALHAVEN REGION

Famous for its white sands, azure water and marine life including dolphins, fur seals and wintertime-whales, Jervis Bay has long been a popular beach camp, with a number of commercial and national-park campgrounds lining its shores. But arguably the loveliest local campground of them all is on the Beecroft Peninsula on the northeastern side of the bay.

This dramatic sandstone plateau has been used by the Australian Defence Force for weapons training activities since the 1800s, and these activities continue on the peninsula today, making it off-limits to visitors on weekdays. But come Friday afternoon,

it opens to campers hoping to snag one of 62 spots in the rustic campground (BYO everything) fringing Honeymoon Bay, where two headlands protect a natural ocean pool. With a ballot system operating during the summer

 **THE PITCH**

It's only open on weekends and school holidays for an unusual reason, but this basic bush campground is still a winner, wrapping around a seriously romantic cove.

school holidays, spontaneous campers would be wise to arrive when the gates open at 1pm on Fridays outside this period, when sites are available on a first-come, first-served basis. Swim, snorkel and discover local walking tracks (staying on the designated path). Or check out the Point Perpendicular Lighthouse on the peninsula's southern tip.

When: year-round
Amenities: toilets, rubbish bins
Best accessed: by car
Cost: $15
Contact: www.shoalhaven.com

HONEYSUCKLE CAMPGROUND
NAMADGI NATIONAL PARK, AUSTRALIAN CAPITAL TERRITORY

22

The Australian Capital Territory might only have one national park, but it's a doozy, protecting 1061 sq km (410 sq miles) of alpine, sub-alpine and mountain bushland covering nearly half of the territory. Of the three campgrounds inside the national park, on the traditional lands of the Ngunnawal people, Honeysuckle Campground is easily the most intriguing. It's located just steps from the former Honeysuckle Creek Tracking Station, which received and relayed to the world the first televised footage of US astronaut Neil Armstrong setting foot on the moon on 20 July, 1969.

The tracking station was decommissioned in 1981, with little more than a concrete slab left behind, but educational signage nonetheless offers a fascinating glimpse into the glory days of the Apollo missions. And it's not the only highlight of staying at the Honeysuckle Campground, which also offers easy access to the epic granite cliffs of Booroomba Rocks as well as bushwalking and rock climbing at Orroral Ridge. The epic Australian Alps Walking Track also weaves past this campground, and while there's a little more light pollution from nearby Canberra these days, the crisp skies above the wheelchair-accessible bush campground are still excellent for stargazing on a moonless night.

🛖 THE PITCH

Stargaze from a peaceful bush campground beside a historic site that once played a pivotal role in the history of space exploration.

When: Oct–Apr
Amenities: toilets, picnic tables, barbecues, firepit
Best accessed: by car
Cost: $11.10
Contact: www.parks.act.gov.au

COTTER CAMPGROUND
COTTER RECREATION AREA, AUSTRALIAN CAPITAL TERRITORY

A favourite weekend getaway for Canberrans, Cotter Campground isn't exactly the kind of place you go to seek solitude under the stars. But when all you want is an easy nature-based escape, it delivers. Stretching along the banks of the peaceful Cotter River where it meets the equally lovely Murrumbidgee River, this large, wheelchair-accessible campground is just 25 minutes from the city centre. And it can accommodate just about every conceivable camping set-up, with hot showers, flushing toilets, potable water and dishwashing facilities a plus for campers partial to creature comforts.

Rise early to try your luck spotting a platypus paddling in the river, or content yourself with the company of kangaroos and emus known to visit the campground. Follow the bends of the Cotter River to the nearby Cotter Avenue parklands, where the Cotter Dam Discovery Trail leads to the Cotter Dam viewing area. Or stretch your legs on the 4.5km (2.8-mile) return Bullen Track, which explores both sides of the river. Yarn with other campers around the communal barbecue area, or relax by the river in the shade of deciduous trees that burst into fiery autumnal hues as winter creeps closer.

THE PITCH

Get back to nature with ease on a leisurely stay at this family-friendly campground on the banks of the Cotter River, just an easy hop from central Canberra.

When: Oct–Apr
Amenities: drinking water, toilets, showers, barbecues, picnic tables, firepit
Best accessed: by car or bus
Cost: $16.70
Contact: www.parks.act. gov.au

23

PORT STEPHENS KOALA SANCTUARY GLAMPING

PORT STEPHENS, NSW

Just north of Newcastle, Port Stephens is home to one of the last remaining koala populations on Australia's east coast. Here the local council has joined forces with volunteer rescue group Port Stephens Koalas to transform a former caravan park behind One Mile Beach into an eight-hectare wildlife haven where sick, injured and orphaned koalas can be rehabilitated for release back into the wild. Opened in 2020, the Port Stephens Koala Sanctuary also offers the chance to bed down alongside its furry guests.

Lining a leafy interpretative pathway linking the visitor centre to the elevated koala enclosure are 20 raised, light-filled glamping tents with air-con and generous decks. At 5pm, when the last day visitors have left, the bushland setting feels even more peaceful. While you can't see (easily, at least) the koalas from your tent, you might hear the bellow of a randy male during mating season. And waking up to native birdsong is a given. Included with all stays is a guided tour of the sanctuary before it opens to the public – and when the koalas are most active. Consider combining a stay with a hike on the newly opened Tomaree Coastal Walk (27km/16.8 miles) that threads past the sanctuary. Every visit helps to fund the sanctuary's work.

24

© CHRIS ANDREWS FERN BAY | SHUTTERSTOCK

THE PITCH

Find a home among the gum trees – if only for one night – as the koalas do at this unique sanctuary welcoming overnight guests.

When: year-round
Amenities: drinking water, private bathroom, kitchenette, electricity, swimming pool, wi-fi
Best accessed: by car or bus
Cost: $475
Contact: www.portstephenskoala sanctuary.com.au

NORTHERN TERRITORY

From its remote red deserts to its teeming tropical wetlands, the Northern Territory's wild and wonderful landscapes were made for outdoor adventures.

Best time: May–Oct (Tropical North), May–Sep (Red Centre)

Best national parks for camping: Kakadu National Park, Litchfield National Park, Nitmiluk National Park, Tjoritja/West MacDonnell National Park

Best camping trails: Larapinta Trail, Jatbula Trail

National parks pass required: Yes, for most parks (NT residents exempt)

Useful contacts: www.northernterritory.com; https://nt.gov.au/parks

Charles Darwin never got to visit the Australian city that was named for him. It's a shame, as the Northern Territory capital is the jumping-off point for some of the world's most thrilling adventures. Whether you head east to Kakadu, Arnhem Land and the Gulf of Carpentaria, north to the Tiwi Islands, or south through Litchfield and Nitmiluk en route to the Red Centre, vast tracts of wilderness heaving with native wildlife beckon.

It's easy to believe that the Northern Territory's extraordinary landscapes were formed by ancient Creator beings. Here, more than 65,000 years of culture continue to pulse through the land. An ever-expanding array of Aboriginal experiences offers memorable opportunities to deepen your connection to the Northern Territory, while the hundreds of trails threading through its national parks immerse you in its natural beauty at your own pace.

Yes, the Territory's harsh climate and native wildlife can be unforgiving – did we mention there's one croc for every 2.5 locals? But with a little preparation, this wild frontier is more accessible than you might think. Note that NT campgrounds are graded A, B, C or D according to their facilities.

FREE CAMPING

Most free camping spots are located along the Stuart Highway – free camping is not permitted within the council areas of Darwin and Palmerston. Bookings are required to camp in national parks.

SUPPLIES

Darwin has the lion's share, including Mitchells (www.mitchellsadventure. com), Camping World (www. campingworlddarwin.com.au), BCF (www.bcf.com.au) and Anaconda (www.anacondastores.com). Collecting firewood in national parks is not allowed. Be mindful that few parks have rubbish bins.

SAFETY

Among the biggest dangers is the heat. Keep an eye on the weather forecast and dress for the conditions. Bring snacks for energy, avoid walks in extreme heat, and if you feel unwell, stop, rest in the shade, and drink water. Be prepared for road closures during the wet season. If you break down at any time, stay with your vehicle. And always be 'crocwise' (www. becrocwise.nt.gov.au).

BEST REGIONS
Kakadu National Park

Roughly the size of Wales, Australia's second-largest national park lures travellers with

Swimming in Kakadu National Park (left) but watch for crocs everywhere (below)

its Aboriginal rock art, serene swimming holes, thriving wetlands and spectacular stone country, with over a dozen memorable spots to spend the night.

Red Centre

Whether it's the rugged red cliffs of the Tjoritja/West MacDonnell Ranges outside Alice Springs, or the blockbuster beauty of Uluṟu-Kata Tjuṯa National Park, Australia's desert heart is a guaranteed showstopper.

Litchfield National Park

Proximity to Darwin combined with lush swimming holes, towering termite mounds and unique geological formations makes Litchfield a cracking place to pitch a tent.

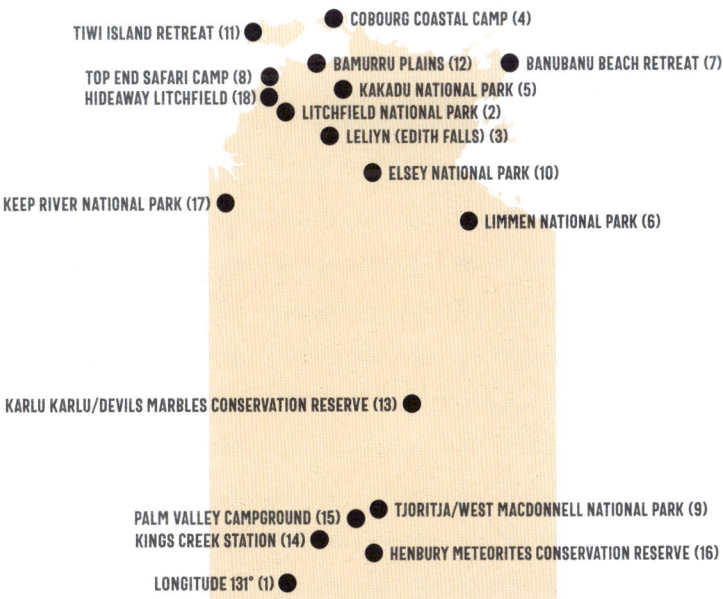

COBOURG COASTAL CAMP (4)

TIWI ISLAND RETREAT (11)

BAMURRU PLAINS (12)   BANUBANU BEACH RETREAT (7)

TOP END SAFARI CAMP (8)   KAKADU NATIONAL PARK (5)
HIDEAWAY LITCHFIELD (18)   LITCHFIELD NATIONAL PARK (2)

LELIYN (EDITH FALLS) (3)

ELSEY NATIONAL PARK (10)

KEEP RIVER NATIONAL PARK (17)

LIMMEN NATIONAL PARK (6)

KARLU KARLU/DEVILS MARBLES CONSERVATION RESERVE (13)

PALM VALLEY CAMPGROUND (15)   TJORITJA/WEST MACDONNELL NATIONAL PARK (9)
KINGS CREEK STATION (14)   HENBURY METEORITES CONSERVATION RESERVE (16)

LONGITUDE 131° (1)

© CHRIS CHEN | LONELY PLANET; TIWI ISLAND RETREAT

LONGITUDE 131°
ULURU, RED CENTRE

The pinnacle of Australian glamping, Longitude 131° elevates a visit to Uluru to an A-list experience. Nestled in rust-red dunes just outside the region's small resort village of Yulara, all 16 of its luxuriously furnished ensuite glamping tents feature mesmerising views of the sacred monolith. You could spend all day gazing at this natural wonder from your king-size bed or private deck (complete with a fireplace to warm your bones on cool winter nights) – but that would be a waste of the slew of options included in the tariff. Venture out to sip premium wines at an exclusive pop-up bar as Uluru changes colour in the fading light, dine under the stars while a guide traces the southerly constellations, and hike around Uluru-Kata Tjuta National Park's famous rock formations, which hold deep cultural significance for the Yankunytjatjara and Pitjantjatjara peoples. Splurge on the dreamy Dune Pavilion, the only tented suite offering dual views of Uluru and Kata Tjuta, as well as an outdoor hot tub. Or simply seize the opportunity to roll out a complimentary swag on your deck and sleep under the Milky Way on a crystal-clear desert night. Transfers from Yulara's Ayres Rock Airport included.

🍽 THE PITCH

Savour uninterrupted views of Australia's most famous rock from your bed at one of the nation's most iconic luxury glamping retreats – the closest accommodation to Uluru.

When: May–Sep
Amenities: drinking water, restaurant, bar, wi-fi, electricity, private bathroom, guest lounge, pool
Best accessed: by car
Cost: from $8400 for 2pp for 2 nights
Contact: www. longitude131.com.au

01

© LONGITUDE 131°

LITCHFIELD NATIONAL PARK
TROPICAL NORTH

Waterfalls cascade into clear pools in monsoonal vine forests and termite mounds rise out of the red earth in Litchfield National Park, a verdant oasis just 100km (62 miles) southeast of Darwin. Covering around 1500 sq km (580 sq miles), the region is important to the Koongurrukun, Mak Mak Marranunggu, Werat and Warray Aboriginal peoples, whose ancestral spirits formed the landscape you see today. As you paddle in limpid water, wander lush, shady trails and explore the Lost City – a cluster of sandstone pillars that resemble an ancient ruin – you'll understand why.

Wangi Falls and Florence Falls are the pick of the national park's camping areas (book ahead). Both Category A campsites with a good range of facilities, these 2WD-accessible sites also lie within walking distance of the park's top swimming holes. There are also 4WD camping areas (dry season only) at Central Valley, Sandy Creek and Surprise Creek Falls, and walk-in campsites at Walker Creek – all with nearby swimming spots. Swimming is riskier in the wet season, when crocs are known to stray into the park, so be sure to check for warnings before you dive in. Grab last-minute supplies in Bachelor, a 45-minute drive away.

🖲 THE PITCH

Camp within stumbling distance of some of the Territory's dreamiest waterfalls and swimming holes, with six campsites to choose from in this surreal desert oasis.

When: May–Oct
Amenities: main sites have untreated water, toilets, showers and firepits or gas barbecues
Best accessed: by car
Cost: $10–20
Contact: https://nt.gov.au/parks

03

LELIYN (EDITH FALLS)
NITMILUK NATIONAL PARK, KATHERINE REGION

The drone of cicadas echoes through the gorges of Nitmiluk National Park, about 320km (200 miles) south of Darwin. Its name means Cicada Place in the language of the Jawoyn peoples who own and manage this rugged wilderness area. There are several private campgrounds closer to the main park entrance, near the town of Katherine, as well as basic campsites in the southern section of the park that can be accessed by canoe or by hiking in on the Southern Walks & Trails network. In the northern section of the park, a 60km (37-mile) drive from Katherine, the Leliyn campground (at the

terminus of the famed Jatbula Trail) scores extra points for being 2WD-accessible, yet still feeling deliciously remote. It has good facilities as well as a kiosk that sells basic food supplies.

Look out for the colourful

Gouldian finch and hooded parrot as you pitch your tent just a short stroll from the large swimming hole at the base of the falls. You can also hike upriver to the lovely Sweetwater Pool (8.6km/5.3 miles return), with has a basic campground with toilets if you're inspired to stay longer. Brace for potential swimming hole closures in the wet season.

🔘 THE PITCH

The ancient sandstone gorges of Nitmiluk National Park shelter pandanus-fringed natural pools that'll tempt you to linger in one of the park's most accessible campsites.

When: May–Oct
Amenities: untreated water, toilets, showers, gas barbecues
Best accessed: by car
Cost: $15
Contact: https://nt.gov.au/parks

© IAN CROCKER | SHUTTERSTOCK

COBOURG COASTAL CAMP
GARIG GUNAK BARLU NATIONAL PARK, COBOURG PENINSULA

You'll be asking yourself what exactly is going bump in the night as you lie in bed listening to the wild soundtrack of the Cobourg Peninsula northwest of Arnhem Land, at the very top of Australia's Top End. Perched on blood-red bauxite cliffs overlooking the deceptively idyllic blues of Port Essington, this off-grid camp offers a touch of luxury in one of Australia's most far-flung landscapes. Here the sunset hour calls for slurping wild oysters prised off the rocks before feasting on freshly caught fish. Raised glamping tents are furnished with proper beds, and you might even spot a croc from the loo.

Operated by Venture North Safaris, the camp is used on the company's fishing charters and multi-day 4WD tours from Darwin. While swimming is off-limits, there are plenty of other things to do, including visiting the intriguing ruins of the failed colonial-era Victoria Settlement, fishing for prized barramundi, and visiting one of Australia's most remote cultural centres to learn about local wildlife and the traditions of the local Iwaidja-speaking Aboriginal peoples. There's also a pair of national parks-run camping areas at Smith Point, which have toilets, showers and a limited supply of bore water.

04

© COBOURG COASTAL CAMP

🍲 THE PITCH

Think Arnhem Land is remote? Wait until you reach the other side. Listen to predators hunting under starry skies on a glamping stay in one of Australia's wildest corners.

When: May-Oct
Amenities: drinking water, toilets, showers, electricity, guest lounge
Best accessed: by car (4WD) or boat
Cost: $3450 for a 4-day tour from Darwin
Contact: www.cobourgcoastalcamp.com.au

KAKADU NATIONAL PARK
TOP END

05

This is the Northern Territory's big one. Covering a staggering 20,000 sq km (7720 sq miles) of the Top End, Kakadu is timeless. Recognised by UNESCO for its outstanding natural value and as a living cultural landscape, this wild and rugged wilderness is also home to the world's oldest living cultures.

Camping is one of the best ways to connect with Country (traditional lands), allowing you to experience the landscape at the enchanting times of dawn and dusk. Kakadu has nearly two dozen parks-operated camping areas with varying facilities to choose from, as well as a couple of commercial campgrounds. In a refreshing change from the usual system (Kakadu is a Commonwealth reserve), you don't need to book to camp at parks-operated sites. Top picks include Maguk Campground, near the oasis-like swimming hole that shares its name, and Merl Campground, near the Ubirr rock art site, a stunning lookout over the Nadab floodplain and the crocodile-infested Cahill's Crossing. Deepen your understanding of Kakadu's cultural significance on one of the Aboriginal-guided tours available in the park, such as the Guluyambi Cultural Cruise on the East Alligator River. Some campgrounds and attractions can be visited by 2WD, but a 4WD is ideal.

© CHRIS CHEN | LONELY PLANET IMAGES

🔵 THE PITCH

The spirituality of this ancient landscape is impossible to ignore as you admire Aboriginal rock art, swim in glasslike waterholes and hike the bush trails surrounding Kakadu's many campsites.

When: May–Oct
Amenities: vary – always bring enough drinking water
Best accessed: by car (4WD)
Cost: $0-15 at parks-run sites
Contact: www.parksaustralia.gov.au/kakadu

LIMMEN NATIONAL PARK
GULF REGION

Travel back in time in Limmen National Park, known for its two Lost Cities (located 60km/37 miles apart by road). These eerie sites formed some 1500 million years ago, when a seabed cracked and split as it was exposed, allowing wind and water to carve the sandstone pillars that rise up from the seasonal floodplain.

With no sealed roads and minimal facilities, this isolated national park, some 275km (171 miles) southeast of Katherine, requires a high level of self-sufficiency. But it's all part of the fun of surrendering to this wild landscape, with a highlight including the rough 28km/17.4-mile 4WD track to the Western Lost City, with its ancient geological features and oodles of birdlife.

There are eight designated camping areas in the Territory's third-largest national park, but two deserve a special mention. The Southern Lost City Campground lies right next to the 2.5km (1.6-mile) walking trail looping around the pinnacles; pitch up on the western side of the loop for a front-row view. A 35km (22-mile) drive north on the Nathan River Road, Butterfly Falls Campground offers the park's only opportunity to swim safely, though this waterhole can become stagnant towards the end of the dry season.

🗨 THE PITCH

Slip off the grid — and the sealed road — in this remote park on the edge of the Gulf of Carpentaria, where the giant sandstone pillars of two 'lost cities' loom large.

When: May–Oct
Amenities: toilets, firepits
Best accessed: by car (4WD)
Cost: $10–15
Contact: https://nt.gov.au/parks

06

Four-wheel driving

Start your engine for adventures of the off-road kind, with some of Australia's most thrilling 4WD tracks taking you through the wildest corners of the Northern Territory.

Bumping along Australia's gnarly 4WD tracks is one of the nation's most popular pastimes, with countless dedicated blogs and YouTube shows, and even a national body dedicated to representing the local 4WD community (www.4wda.com.au).

While you'll need a 4WD to access many remote areas and national parks in the Northern Territory and beyond, four-wheel driving is considered among purists to refer to technical tracks that put both your driving skills and patience to the test.

VEHICLE HIRE

4WD hire is available in Darwin and Alice Springs, with some companies offering pick-up or drop-off in Cairns, Broome, and even Perth. Vehicles kitted out with camping equipment cost from around $300 per day, usually with the addition of a hefty insurance bond; book well in advance. Toyota 4WD vehicles are a popular choice; they're reliable and relatively quick to fix as most mechanics stock parts.

SAFETY

It's always wise to travel with a recovery kit (which can be picked up at 4WD Supercentre outlets; www.4wdsupacentre.com.au), and never alone, when leaving Australia's sealed roads, and ideally in a convoy of at least two cars on technical tracks. Four-wheel-driving and sand-driving experience is also a must on these tracks. Carry extra fuel and pick up

a tyre deflator/inflator if you're sand-bound, and if you break down, activate your personal locator beacon and stay with your vehicle until help arrives.

WHEN TO GO

Dusty tracks can turn to mud and swelling rivers can cut off roads during the Top End's wet season (Nov–Apr), while the summer heat can be unbearable in the Red Centre. In other words, aim for the dry season (May–Oct).

Driving in Kakadu National Park (top) and near Darwin (above); Jim Jim Falls in Kakadu (left); watch for kangaroos crossing the road (inset)

FIVE TO TRY

Red Centre Way
Only one section of this 658km (409-mile) drive from Alice Springs to Uluru is unsealed; allow six days.

Finke River 4WD Route
Follow the Finke River from Hermannsburg to Watarrka National Park, camping in Finke Gorge National Park.

Jim Jim Falls
A 4WD is essential to reach this blockbuster Kakadu waterfall, 55km (34 miles) off the sealed road.

Gunbarrel Highway
A formidable 4WD desert track linking Wiluna in WA to Yulara in the NT; allow at least four days.

Binns Track
The 2230km (1386-mile), 10-day epic takes in surreal scenery between the SA border and Timber Creek.

BANUBANU BEACH RETREAT
BREMER ISLAND, EAST ARNHEM LAND

Absorb the natural splendour of East Arnhem Land in style at Banubanu Beach Retreat on Bremer Island, around 1000km (621 miles) northeast of Darwin. Built in partnership with Yolŋu people, the eco-conscious luxury retreat's six spacious glamping tents (including an elevated 'penthouse') dot the sand dunes of this remote island just 5km (3 miles) from the mainland

town of Nhulunbuy, each with sweeping Arafura Sea views. By day, take part in Yolŋu cultural experiences, reel in a prized catch on a fishing charter, swim in tropical turquoise waters, spot some of the 65 bird species recorded in the area or chill out on your private sun deck with a good book. Then, as the sun sets over the ocean, feast on local wild-caught barramundi at

the onsite restaurant. After dark, you might be lucky enough to observe some of the four species of marine turtle that nest on the island's beaches throughout the year before nodding off to the gentle sounds of the sea.

You can drive your own car, plane or boat to Nhulunbuy, 710km (441 miles) north from Katherine; or take a scheduled flight to its Gove-East Arnhem Airport from Darwin or Cairns. Retreat rates are all-inclusive.

07

🪙 THE PITCH
Glamp on the edge of the earth — at least, that's what it feels like on Bremer Island, where nothing lies between you and Papua New Guinea except the Arafura Sea.

When: Apr–Nov
Amenities: drinking water, restaurant, bar, non-motorised sports equipment, plunge pool, electricity
Best accessed: by plane, then boat
Cost: from $3948 for 2pp for 2 nights
Contact: www.banubanu. com

08

TOP END SAFARI CAMP

BYNOE

The contrasting landscapes of the Northern Territory dramatically collide southwest of Darwin, where the lush Finnis River system meanders through a sweeping floodplain dotted with termite mounds reminiscent of nearby Litchfield National Park. Straddling these two worlds is Top End Safari Camp. With corrugated iron water tanks and outdoor showers attached to each of its 15 spacious glamping tents, it looks like a classic Aussie outback safari camp. But this remote private option lies just 2km (1.2 miles) from the Finnis River, where exhilarating airboat cruises showcase both the natural beauty and lurking dangers of the river system – including some of the state's largest saltwater crocodiles. Packages include an airboat cruise and a 10-minute scenic helicopter ride over the spectacular floodplain ecosystem, an encounter with a rescued monster crocodile and a barbecue dinner, followed by stargazing around the fire pit. After a restful night's sleep in your tent, which come furnished with double beds and private deck, a hearty cooked breakfast is served to fuel you up for your onward journey. And if the natural air-con doesn't cool you down, you can always jump in the pool.

© TOP END SAFARI CAMP

🍴 THE PITCH

Termite mounds rise up between the deluxe Lotus Belle tents at this outback safari camp, providing a comfortable base after a day of action-packed adventures on the Finnis River.

When: Apr–Oct
Amenities: drinking water, restaurant, pool, electricity
Best accessed: by car
Cost: $1390 for 2pp for 2 nights
Contact: www. topendsafaricamp.com.au

09

TJORITJA/WEST MACDONNELL NATIONAL PARK
RED CENTRE

It's not difficult to picture the giant caterpillars of the Arrernte Dreaming inching across the rugged red landscape west of Alice Springs, forming the dramatic MacDonnell Ranges. Stretching for 161km (100 miles), the cool, scenic gorges of the 'West Macs' provide important refuges for native plants and animals, many of which are found only here. Some are even relics of ancient tropical forests.

Of the park's five main campgrounds, privately managed Ormiston Gorge offers the most facilities, including showers and flush toilets. Ellery Creek Big Hole and Redbank Gorge have toilets and barbecues, while Serpentine Chalet and 2-Mile (4WD only) have no facilities. There are also 41 basic campgrounds along the multi-day Larapinta Trail, which traces the length of the park. Some have untreated tap water, but all campers are advised to bring water supplies with them.

A stroll away from freshwater swimming holes, Ormiston Gorge and Ellery Creek Hole are the pick of the campgrounds. But with no light pollution, the night sky dazzles wherever you pitch your tent.

🖊 THE PITCH

Cool off in a freshwater swimming hole after a day's hiking in this ancient desert landscape near Alice Springs, before bedding down under a million stars.

When: Apr-Sep
Amenities: vary; main campgrounds have toilets and gas barbecues
Best accessed: by car
Cost: $10 for national parks-run sites
Contact: https://nt.gov.au/parks

ELSEY NATIONAL PARK
MATARANKA

Who needs a camp shower when you can roll out of your tent and into a hot spring? Framed by pandanus and paperbark trees, Bitter Springs is the loveliest of Elsey National Park's spring-fed thermal pools. Drift along in the gentle current of its aquamarine waters, a comfortable 34°C year-round. If you packed a snorkel, look out for the freshwater turtles often seen swimming around below you. Nearby Mataranka Thermal Pool is the same temperature, while Stevie's Hole offers the option of a refreshing dip in a cool freshwater pool.

The setting of Jeannie Gunn's classic Australian novel, *We of the Never Never*, the national park – 120km (74.6 miles) south of Katherine – is also rich in pastoral history. Just outside the park boundary, the historic Mataranka Homestead offers camping and rooms within walking distance of Mataranka Thermal Pool. Alongside the Roper River (where swimming is prohibited due to crocs) the well-equipped Jalmurark Campground is the only place to spend the night inside the park, with hot showers and walking trails to the park's swimming spots and historic sites. The springs are typically closed during the wet season due to flooding.

10

TIWI ISLAND RETREAT
BATHURST ISLAND, TIWI ISLANDS

Kick off your shoes, switch off your phone, and open your heart to the Tiwi Islands. A 2.5-hour ferry ride from Darwin, this remote cluster of islands is home to an Indigenous community just as well known for its vibrant textiles as its passion for Australian Rules football. With only a couple of local accommodation options on the two main inhabited islands, Bathurst and Melville, most people visit on a day trip from Darwin with SeaLink ferries. For a special treat, pack your bag for a longer sojourn at Tiwi Island Retreat. On the remote west coast of Bathurst Island, an hour's drive from the main settlement of Wurrumiyanga, this luxury glamping retreat is a scenic and stylish base in which to unplug (there's no mobile reception or wi-fi) and embrace activities from world-class fishing adventures to Indigenous culture tours and unique wildlife encounters.

You can even take a helicopter ride to an idyllic freshwater swimming hole, surrounded by tropical greenery. Cool off after a day of adventures with a refreshing dip in the pool (ocean swimming is off-limits due to saltwater crocodiles) and ease into the evening with canapes served around a beach bonfire as the sun sinks over the Timor Sea.

🐚 THE PITCH

Set sail from Darwin to the wild and lush Tiwi Islands. On the remote west coast of Bathurst Island, this luxe glamping retreat offers a launchpad for adventure and cultural immersion.

Best time: Apr–mid-Dec
Amenities: all-inclusive
Best accessed: by ferry, then car
Cost: from $4840 for 2pp for 2 nights
Contact: www.tiwiislandretreat.com.au

© TIWI ISLAND RETREAT

BAMURRU PLAINS
MARY RIVER

Feel your adrenalin levels surge as your wheels leave the tarmac partway into the three-hour journey north from Darwin to 'Australia's Okavango Delta'. The unsealed road brings you to Bamurru Plains, a luxe and ultra-eco-friendly off-grid safari camp with just 12 lodgings sharing exclusive access to 300 sq km (116 sq miles) of floodplains and savanna woodland on the Mary River, on the border of Kakadu National Park.

This wild, tropical landscape was once a significant meeting place for several Aboriginal groups – 'bamurru' is a Gagadju word for the magpie geese that flock to this fertile coastal floodplain throughout the year. They're just one of 236 bird species you're likely to spot on included twice-daily safari-style excursions that immerse visitors in this teeming tropical wilderness, home to the world's highest concentration of estuarine crocodiles.

This could mean zooming across the watery landscape on an airboat, bumping along in wonder on open-top safari drives, eyeballing crocs on river cruises, or venturing deep into the bush on guided bush walks. Feel at one with nature even in bed in your glamping tent, which feature three walls of floor-to-ceiling mesh screen, as water buffalo amble by in silhouette against the moon.

12

© BAMURRU PLAINS

🐍 THE PITCH

Embark on an Australian safari at this tropical Top End glamping retreat, where native wildlife thrives and Aboriginal history and culture pulse through the land.

Best time: Mar-Oct
Amenities: drinking water, restaurant, private bathroom, electricity
Best accessed: by car
Cost: $5160 for 2pp for 2 nights
Contact: www.bamurruplains.com

Aboriginal Rock Art

With more than 5000 significant rock art sites located in Kakadu alone, the Northern Territory is a great place to admire the artistic and cultural legacy of Australia's First Nations.

A vital part of Australia's Indigenous cultures, rock art offers an intriguing window into how humans have lived and thought on this continent since the earliest evidence of the art form, some 40,000 years ago.

Aboriginal rock art typically takes the form of paintings or petroglyphs depicting everything from now-extinct wildlife to Creation (Dreaming) beings, and is found in every state and territory of Australia except Tasmania, where the function of Aboriginal rock markings remains unknown.

Western Australia lays claim to Australia's oldest rock art painting (a 17,500-year-old kangaroo) as well as the world's largest concentration of rock art at Murujuga, where petroglyphs cover the landscape of the Burrup Peninsula. But it's the richly decorated rock shelves of Kakadu and Arnhem Land in the Northern Territory that steal the spotlight. Not only are there plenty of well-preserved works to be uncovered here, but the x-ray art style traditionally used by the Aboriginal artists of this region depicts subjects in incredible anatomical detail, offering insights into the artists' connections to their Country (traditional lands) and its wildlife.

LOCAL KNOWLEDGE

Interpretative signage at the Northern Territory's largest and most accessible rock art sites helps to contextualise the significance of the works before you, but these paintings are layered with enough stories to fill a million signboards. Join a Bininj (Aboriginal) ranger-guided tour of Ubirr or Nourlangie (Burrungkuy) to connect more

deeply with the artistic legacy of your guide's ancestors; book via www. parksaustralia.gov.au. Possible to visit on a day trip from Jabiru in Kakadu, West Arnhem Land's captivating Injalak Hill rock art site can only be experienced on an Indigenous-guided tour; visit www.injalak.com to book through a partner operator.

ETIQUETTE

It's a basic courtesy to ask before filming or taking photos of a person, a group of people or cultural ceremonies. This courtesy should always be extended not only to Indigenous peoples but also to Aboriginal rock art sites. Each Indigenous group has its own protocols regarding rock art, which may include a photo ban. If you are visiting a rock art site independently, interpretative signage usually notes photography guidelines.

Aboriginal rock art in Arnhem Land (top) and at Ubirr in Kakadu National Park (above); a ranger and rock art (inset) at Nourlangie (Burrungkuy)

FIVE TO TRY

_____

Ubirr,
Kakadu National Park
Rock shelter art depicting fish, turtles, goanna and other core food animals.

Nourlangie (Burrungkuy),
Kakadu National Park
Images of Creation beings, Namarrkon (Lightning Man), European ships and more.

Injalak Hill,
West Arnhem Land
Heavily decorated shelters near Jabiru tell the story of the Kunwinjku people.

Mala Walk, Uluṟu
Wheelchair-accessible walk featuring rock-art sites that tell Aṉangu stories of Tjukurpa (Creation).

Nitmiluk National Park
Jawoyn art can be seen along the base of the sandstone escarpment in the gorge system.

KARLU KARLU/ DEVILS MARBLES CONSERVATION RESERVE
TENNANT CREEK AREA

One of Australia's great outback road trips, the 3000km (1864-mile) Stuart Highway links Darwin to Port Augusta in South Australia. Among the many highlights of this route is this cluster of huge red boulders spread across the landscape about an hour's drive south of Tennant Creek, crudely described by Scottish Royal Navy officer John Ross in 1870 as a bag of marbles emptied by the devil.

Just like the territory's most famous rock – Uluru – Karlu Karlu (which translates to 'round boulders') is particularly moving at sunrise and sunset, when the red granite rocks glow scarlet.

A national park campground just off the highway is suitable for 2WDs, tents and caravans. Several walking tracks criss-cross the site.

Karlu Karlu has been a sacred place for its Kaytete, Warumungu, Warlpiri and Alyawarra Traditional Custodians for hundreds of years. Traditional Custodians ask that visitors do not climb the karlu (marbles).

🔵 THE PITCH

Sacred ceremonies have been conducted at this remote geological wonder for centuries. Today, you can camp amid its unique rock formations with the blessing of the site's Traditional Custodians.

Best time: Apr–Sep
Amenities: toilets, firepits, picnic tables
Best accessed: by car
Cost: $10
Contact: https://nt.gov.au/parks

KINGS CREEK STATION
RED CENTRE

On the doorstep of the magnificent Watarrka National Park, Kings Creek Station is a working cattle property established by the son of the region's first permanent pastoralist. In recent years, tourism has become an increasingly important part of the business, with five camping options ranging from unpowered bush camping among majestic desert oaks to deluxe Dreamtime Escarpment glamping tents nestling alongside the dramatic George Gill Range. You can tour the 2200 sq km (849 sq miles) station by buggy or even helicopter, but the best part about staying here is the easy access to Kings Canyon, the jewel of Watarrka National Park, just 36km (22 miles) down the road. Seize the chance to lap up the scenery from its sunset viewing area after the tour buses from Alice Springs and Uluru have left for the day.

One of only two commercial campgrounds within easy striking distance of the park (there's also a walk-in campground inside the park), Kings Creek Station is also a five-minute drive from the bush office of Karrke Aboriginal Cultural Experience & Tours, whose one-hour Aboriginal Cultural Tour offers the ultimate introduction to the significance of this landscape to Luritja and Arrernte peoples.

🗨 THE PITCH

Get a taste of life on an outback cattle station on an overnight stay, with the bonus of beating the crowds to nearby Watarrka (Kings Canyon) National Park in the morning.

Best time: May-Oct
Amenities: untreated water, toilets, showers, pool, laundry facilities, restaurant, electricity
Best accessed: by car
Cost: camping from $27.50
Contact: www.kings creekstation.com.au

14

© KINGS CREEK STATION

15

PALM VALLEY CAMPGROUND
FINKE GORGE NATIONAL PARK, RED CENTRE

Follow the track along the dry, sandy bed of the Finke River (in a high-clearance 4WD only) from Hermannsburg for 18km (11 miles) to the premier campground in Finke Gorge National Park. Here the impressive Palm Valley, which holds cultural significance for Western Aranda people, is home to many rare plants including a 12,000-strong population of red cabbage palms found nowhere else on Earth. This remnant of Central Australia's tropical past, 138km (86 miles) west of Alice Springs, was a key inspiration for watercolourist Albert Namatjira (1902–1959), an Arrernte man from the West MacDonnell Ranges

and a pioneer of contemporary Indigenous Australian art.

Adjacent to Palm Creek, the Palm Valley Campground is 4km (2.5 miles) east of Palm Valley. Shaded by river red gum trees, it has good facilities and sites

available for tents, camper trailers and even caravans, if you're game to tow one here. From the Palm Valley car park, the Arankaia Walk (2km/1.2 miles, 1hr) and the longer Mpulungkinya Walk (5km/3.1 miles, 2hrs return) meander through a lush landscape of slender palms and return to the car park across the plateau. Two more walks begin at the Kalarranga car park.

⊕ THE PITCH

Feel like you've stepped into a real-life watercolour painting in this pocket of palms framed by scarlet cliffs, just down the creek from a similarly scenic national park campground.

Best time: May–Sep
Amenities: untreated water, toilets, hot showers, barbecues
Best accessed: by car (4WD)
Cost: $15
Contact: https://nt.gov.au/parks

HENBURY METEORITES CONSERVATION RESERVE
RED CENTRE/ALICE SPRINGS REGION

Some 4700 years ago, the Henbury Meteorite hurtled toward Earth, fragmenting before impact in the Central Australian desert, 145km (90 miles) southwest of Alice Springs. It's any wonder what was going through the minds of the region's First Peoples who witnessed the spectacle, which continues to feature in the stories and traditions of the local Luritja people.

The dozen craters left behind now form part of the Henbury Meteorites Conservation Reserve, just 13km (8 miles) from the Stuart Highway via a 2WD-accessible gravel road. An easy 1.5km (0.9-mile) loop walking trail from the basic campground connects the main craters, the largest measuring 180m (591ft) wide and 15m (49ft) deep. You'll need to look a bit harder for the smallest, a shallow indentation just 6m (20ft) wide. While there isn't much else to see in this flat, sparsely vegetated landscape, camping here offers an opportunity to savour the mystical appeal of watching the fading light play on the craters at sunset.

Some 500kg (1100lbs) of octahedrite meteorite, consisting of 90 per cent iron and 8 per cent nickel, have been recovered from the site to date. If you're headed onward to Alice Springs, you can see some of the fragments at the Museum of Central Australia.

16

© LKPRO | SHUTTERSTOCK

THE PITCH

Watch the moon rise over an alien landscape on a night camping beside one of the world's best preserved small crater fields, in the middle of the Australian outback.

Best time: May-Sep
Amenities: toilets, firepits, picnic tables
Best accessed: by car
Cost: $10
Contact: https://nt.gov.au/parks

KEEP RIVER NATIONAL PARK
SOUTHWEST OF DARWIN

Remote Keep River National Park nestles alongside the Western Australian border, with Kununurra the closest town, 46km (28.6 miles) to the west. An important site to the Miriwoong and Gajirrabeng peoples for thousands of years, this relatively small park packs some seriously epic scenery into its 574 sq km (222 sq miles), with Aboriginal rock art and surreal sandstone formations reminiscent of the Bungle Bungle Range.

There are two basic, shady camping areas – Goorrandalng, 18km (11 miles) from the park entrance, and Jarnem, 32km (20 miles) from the entrance. Accessible by 2WD in the dry season, both are worthy options. From Goorrandalng, which allows generators, a 2km (1.2-mile) loop trail weaves through a soul-stirring sandstone environment. From Jarnem, which has limited untreated tap water (tap water is also available near the park entrance, just past the ranger station), a 7km/4.3-mile loop track takes you to a rock art site as well as a spectacular lookout. Other short walks leading off the main access road reveal fascinating traces of historic Aboriginal occupation. Keep an eye out for the short-eared rock-wallaby, white-quilled rock pigeon and sandstone shrike-thrush.

17

HIDEAWAY LITCHFIELD
RAKULA

18

When you're looking for a stylish, 2WD-accessible Northern Territory wilderness retreat that won't break the bank, this trio of modern cabins nestled in the tropical bushland of the Litchfield region delivers. On a private property just outside Litchfield National Park, each of its one-bedroom cabins are made from 12m (40ft) shipping containers sliced in half – producing two double-storey spaces (Tolmer and Wangi) and one single-level pavilion (Cascade), ideal for guests with less mobility. Each of the villas has a lounge area, fully equipped kitchen, bathrooms with rain showers and a deck with a barbecue, perfect for frying up the breakfast provisions supplied in your fridge. Usually there's a two-night minimum, but if there's a single night available between bookings, you're welcome to snap it up. Children (or an extra adult) can be accommodated.

Just 8km (5 miles) from Wangi Falls, Hideaway Litchfield is ideally situated for exploring its eponymous national park. Yet with their huge bedroom windows inviting the outside in (minus the critters), the cabins, situated 70m (230ft) apart, offer a nature immersion of their own. Newer to the property are a trio of cosy queen-bed huts, with cheaper rates but less privacy, as they are clustered together.

🔖 THE PITCH

The Northern Territory's only modern wilderness cabin accommodation offers an elegant nature-based stay for a more affordable price tag than a high-end glamping retreat.

Best time: May–Oct
Amenities: private bathroom, electricity, kitchen, wi-fi, barbecue
Best accessed: by car
Cost: from $420 per cabin
Contact: www.hideawaylitchfield.com

© JASON HILL

QUEENSLAND

Fringed by coral reefs, blanketed in lush rainforest, dotted with outback oases and bursting with wildlife, the Sunshine State isn't short on spots to escape the urban grind.

Best time: May–Oct (tropical north, outback); year-round (southeast Queensland)

Best national parks for camping: Gheebulum Kunungai (Moreton Island), Girraween, K'gari, Whitsunday Islands

Best camping trails: Thorsborne Trail, Scenic Rim Trail, K'gari Great Walk, Carnarvon Great Walk

National parks pass required: No, only for driving in recreation areas on Bribie Island, K'gari and Cooloola

Useful contacts: www.queensland.com; https://parks.des.qld.gov.au

With more than 6.5 million hectares of national parks alone to explore in Queensland, where does one begin? If you love an ancient rainforest, there's the Gondwana Rainforests in the south, and the Daintree Rainforest in the north – both of them World Heritage-listed. And there are plenty of other lush naturescapes in between. More of a beach person? Pitch up along the state's 7000km (4300 miles) of coastline, and that's just on the mainland. Offshore islands beckon the ultimate castaway adventure, while a vast outback reveals geothermal springs, unexpected bird sanctuaries, rolling red dunes and other surprises.

Queensland's year-round warm weather was made for sleeping alfresco, with plenty of adventures to occupy you in the state's southeast when the summertime heat and wet-season (Nov–Apr) rains make it more difficult to explore Queensland's outback and tropical north.

FREE CAMPING

While you'll need to book and pay to camp in all Queensland national parks, it's a bargain flat rate of $7.25 per person, per night. There are also plenty of free camping sites tucked away around the state. Babinda Boulders near Cairns and Toomulla Beach Campground north of Townsville are scenic choices.

SUPPLIES

Queenslanders love the outdoors, and you'll see this reflected in the abundance of camping and fishing stores in every city and town of size. The Gold Coast, Brisbane/Meeanjin, Townsville and Cairns/Gimuy offer the most variety.

SAFETY

Croc country begins north of the Bundaberg region. In recent years stingers have been recorded as far south as K'gari during the season (Nov–May), but the danger zone is widely considered to be north of Gladstone. Queensland's vast outback commands considered preparation, and dingo safety is paramount on K'gari.

BEST REGIONS

Wet Tropics

Spending a night or five immersed in the steamy rainforests of tropical north Queensland as fireflies dance on the breeze and lace monitors prowl the forest floor is something else. There are more than two dozen national

Pack snorkels and fins for Wilson Island (left); a forest dragon in the Daintree rainforest

parks in the Wet Tropics World Heritage Area, and you can camp in most of them.

Great Barrier Reef
Far from just a day-trip destination, the world's largest coral reef system has some incredible camping and glamping options on islands and pontoons moored in protected lagoons.

Southeast Queensland
From ancient Gondwana Rainforests to the high-altitude bushland of the Granite Belt, the state's southeast corner is home to a surprising diversity of landscapes, with mild winters and the lack of a wet season making it a great region to camp year-round.

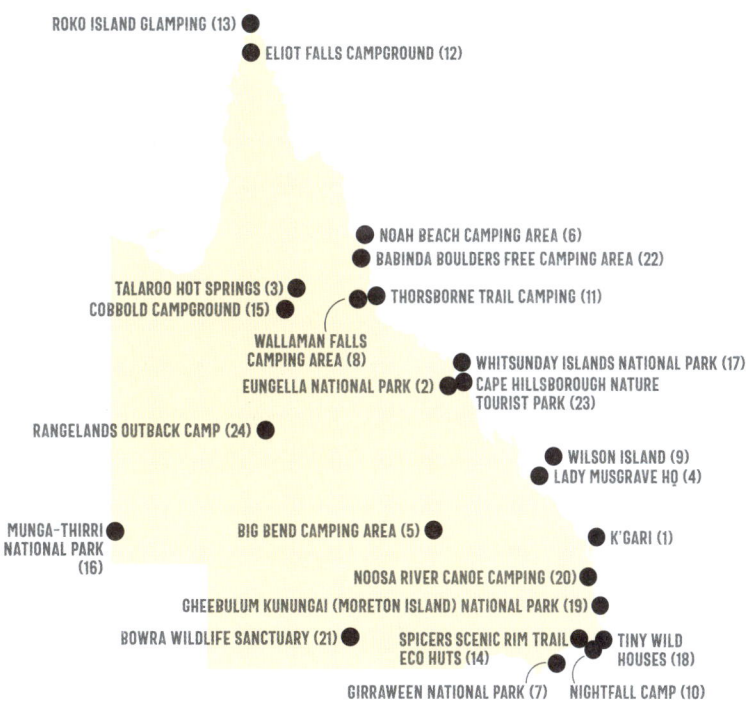

ROKO ISLAND GLAMPING (13)

ELIOT FALLS CAMPGROUND (12)

NOAH BEACH CAMPING AREA (6)
BABINDA BOULDERS FREE CAMPING AREA (22)

TALAROO HOT SPRINGS (3)
COBBOLD CAMPGROUND (15)

THORSBORNE TRAIL CAMPING (11)

WALLAMAN FALLS CAMPING AREA (8)

WHITSUNDAY ISLANDS NATIONAL PARK (17)

EUNGELLA NATIONAL PARK (2)
CAPE HILLSBOROUGH NATURE TOURIST PARK (23)

RANGELANDS OUTBACK CAMP (24)

WILSON ISLAND (9)
LADY MUSGRAVE HQ (4)

MUNGA-THIRRI NATIONAL PARK (16)

BIG BEND CAMPING AREA (5)

K'GARI (1)

NOOSA RIVER CANOE CAMPING (20)

GHEEBULUM KUNUNGAI (MORETON ISLAND) NATIONAL PARK (19)

BOWRA WILDLIFE SANCTUARY (21)

SPICERS SCENIC RIM TRAIL ECO HUTS (14)

TINY WILD HOUSES (18)

GIRRAWEEN NATIONAL PARK (7)
NIGHTFALL CAMP (10)

K'GARI
GREAT SANDY NATIONAL PARK, FRASER COAST

K'gari lies little more than a kilometre off the coast of southeast Queensland at its closest point, but this sandy wilderness (formerly known as Fraser Island) may as well be a world away. Here, on the traditional lands of the Butchulla people, majestic remnants of tall rainforest rise out of the world's longest and most complete age sequence of coastal dune systems, and half of the world's perched freshwater dune lakes (including the neon-blue Lake McKenzie) offer magical opportunities to cool off.

Access to the island is by barge from River Heads in Hervey Bay, or Inskip Point at Rainbow Beach, from where Pacific coastline stretches 122km (76 miles) to Sandy Cape. With the absence of sealed roads making driving a laborious affair, you'll want to spend at least a few days – six-to-eight if you're planning to tackle the K'gari Great Walk (90km/56 miles). Scope out your own private spot within nine camping zones (with no facilities) or choose from an additional 25 campgrounds with facilities ranging from none to sites with firepits, flushing toilets, hot showers and dingo fences. There are five places to fuel up on K'gari, each with basic supplies, but it's best to bring everything you'll need for your adventure with you.

💡 THE PITCH

Hear the howls of dingoes carry on the ocean breeze as you stake out your own private beach-camping spot in this World Heritage-listed wilderness.

When: year-round
Amenities: varies
Best accessed: by car (4WD only)
Cost: $7.25
Contact: https://parks.des.qld.gov.au

01

© LOUIELEA | SHUTTERSTOCK

EUNGELLA NATIONAL PARK
MACKAY HIGHLANDS

High above the sugar cane plantations, the misty mountain refuge of Eungella National Park is one of Queensland's most ecologically diverse protected places, home to 860 plant species alone. But most visitors come to see its healthy population of platypuses. On opposite sides of the slow-moving Broken River, the park's two campgrounds position you right in the action.

At the entrance to the Broken River section of the park, next to the Broken River Bridge, the Broken River Camping Area has eight open-plan campsites (suitable for caravans) set amongst natural bushland. Individual sites

can't be booked ahead, so you'll want to get here early to nab a riverbank spot. Across the bridge is the peaceful, grassy Broken River Visitor Area, with barbecues, picnic tables, a great platypus viewing area and access to a handful of

rainforest trails. Just 600m (1968ft) upriver, the Fern Flat Camping Area nestles in denser vegetation, suitable for tents and campervans only. Both campsites offer great vantage points to spot the quirky monotremes, particularly at dawn and dusk. Bring warm clothes to wear while you wait for the magic to unfold – it can get chilly at an elevation of 700m (2296ft).

🪙 THE PITCH

Observe one of Australia's most elusive animals in its natural habitat from your tent at a choice of two elevated subtropical rainforest campsites on the banks of a languid river.

When: year-round
Amenities: toilets, plus untreated water at Fern Flat Camping Area
Best accessed: by car
Cost: $7.25
Contact: https://parks.des.qld.gov.au

TALAROO HOT SPRINGS
GULF SAVANNAH

In the heart of Gulf Savannah country, scorching water bubbles from the earth, spilling into terraces and lighting up the semi-arid landscape in a riot of colour as the mineral-rich water interacts with land-side microorganisms.

Dispossessed from this ecologically and culturally significant area during colonisation, the Ewamian people regained partial ownership of their traditional homelands in 2012 when the 31,500-hectare Talaroo Station was purchased on their behalf. In 2021 they opened its most spectacular natural feature to the public, along with an excellent campground.

Just 10km (6.2 miles) off the epic Savannah Way touring route via a 2WD-accessible dirt road, about 4.5 hours west of Cairns, the campground features 39 spacious sites dotted with trees, though don't expect much shade. There are also two privately situated 'eco tents' where termite mounds dot the horizon beyond a private deck. A raised boardwalk snakes above the steaming springs. Visits, by guided tour only, include a dip in a spring-fed soaking pool. Private soaking pools can be booked, and a new bike trail that begins at the campground features bush tucker, birdwatching opportunities and cultural sites including scar trees.

THE PITCH

Soak away your worries in the healing waters of an ancient outback spring before floating back to your tent in the Aboriginal-owned campground.

When: Apr–Sep
Amenities: drinking water, camp kitchen, electricity, wi-fi, waste facilities, swimming pool
Best accessed: by car
Cost: from $32
Contact: www.talaroo.com.au

03

LADY MUSGRAVE HQ
SOUTHERN GREAT BARRIER REEF

It's an exhilarating feeling to be marooned on the Great Barrier Reef — on purpose, that is. As the *Reef Empress* motors back to Bundaberg laden with day trippers, the southern fringe of the UNESCO-listed reef system suddenly feels more alive. More remote. More mysterious. This is your chance to slip into the protected coral lagoon surrounding uninhabited Lady Musgrave Island for a serene snorkelling session alongside a maximum 15 fellow overnight guests — and a couple of million marine critters.

This is just the beginning of a memorable night glamping in one of eight queen tents on the top floor of the Lady Musgrave HQ, a permanent, eco-sensitive pontoon operated by Bundaberg-based operator Lady Musgrave Experience. Enjoy a hearty evening meal as the sun sets behind the reef, and a canopy of stars unfurls above your bed. From October to January, there's also an opportunity to take a tender to Lady Musgrave Island to watch marine turtles nesting in the moonlight. You can alternatively (or additionally) BYO everything you'll need for a few nights camping at the island's rustic, national parks-run campground for the usual $7.25 per person per night rate before hitching your way back to Bundaberg on the *Reef Empress*.

04

🔆 THE PITCH

Experience the Great Barrier Reef in a thrilling new way on a glamping experience that immerses you in the drama of the reef — which doesn't stop when the sun sets.

When: year-round
Amenities: drinking water, toilets, showers, electricity, restaurant
Best accessed: by boat
Cost: $1250 including transfers and meals
Contact: www.ladymusgrave experience.com.au

BIG BEND CAMPING AREA
CARNARVON NATIONAL PARK, CENTRAL QUEENSLAND

A hidden sanctuary in the semi-arid heart of Central Queensland, the sandstone cliffs of Carnarvon Gorge cradle a wealth of cultural and natural heritage. Here endemic plants including ancient cycads and Carnarvon fan palms line the main gorge, remnant rainforest flourishes in sheltered side-gorges and grassy open forest clings to the cliff tops. The boulder-strewn Carnarvon Creek gurgles through the natural chasm, attracting wildlife and birds galore, while ochre stencils, rock engravings and freehand paintings on sandstone overhangs at the Art Gallery and Cathedral Cave are reminders of Aboriginal peoples' long and continuing connection with the landscape. The main gorge walking track takes it all in on its winding, 9.7km (6-mile) route from the visitor centre to the Big Bend Camping Area, with side tracks en route leading to attractions including the fairy-tale-like Moss Garden and an impressive natural amphitheatre.

While the main Carnarvon Gorge Camping Area near the park entrance only opens during the Easter, June-July and September-October Queensland school holidays, the more intimate Big Bend Camping Area is open year-round. It's one of five walkers' camps on the challenging Carnarvon Great Walk (87km/54 miles), which makes a loop around this superb sandstone landscape.

05

🔖 THE PITCH

Admire endemic plants, ancient rock art and wildlife including five of Australia's six species of marsupial glider on a hike to this creekside campsite deep in Queensland's most famous gorge.

When: year-round
Amenities: toilets and untreated water, plus barbecues and picnic tables at the Carnarvon Gorge Camping Area
Best accessed: by car, then on foot
Cost: $7.25
Contact: https://parks.des.qld.gov.au

06

NOAH BEACH CAMPING AREA
DAINTREE NATIONAL PARK, CAPE TRIBULATION

Driving off the Daintree River ferry onto Cape Tribulation feels like entering another realm. Here, in the heart of the world's oldest surviving tropical lowland rainforest, a tangle of greenery cocoons the only access road, creating a mystical portal to a natural wonderland where electric-blue Ulysses butterflies flit between fan palms and dinosaur-like cassowaries could cross your path at any moment. Just off the sealed road ending at the Cape Tribulation headland known as Kulki to the region's Kuku Yalanji Traditional Custodians, Noah Beach Campground is the only place to stay inside the park.

Pitch a tent or park your camper trailer under the rainforest canopy, protected from the ocean breeze and tropical sun. Watch fireflies dance in the dusk and wake to the sound of the Coral Sea lapping at the shore – Noah Beach is just 50m (164ft) away. Lurking salties (saltwater crocodiles) mean it's not safe to swim in the sea or in nearby Noah Creek, but there are several freshwater swimming holes within a short drive. There are also several privately-run campgrounds in Cape Tribulation with more facilities, outside the national park boundary but also surrounded by rainforest.

💬 THE PITCH

Pitch a tent at the convergence of two World Heritage Sites north of Cairns, where crocs lurk, cassowaries roam, and insect repellent is always a good idea.

When: May–Oct
Amenities: toilets
Best accessed: by car
Cost: $7.25
Contact: https://parks.des.qld.gov.au

GIRRAWEEN NATIONAL PARK
GRANITE BELT

Pack warm clothing (even in summer) and take an easy country drive 260km (162 miles) southwest of Brisbane to Girraween National Park on the Queensland–New South Wales border. Girraween is an Aboriginal word meaning 'place of flowers', and if you visit in spring, you'll be rewarded with bursts of wildflowers blanketing the landscape. But there is plenty more to enjoy in this national park year-round, where giant boulders balance precariously on rocky outcrops, cool steams gush into inviting rock pools, and more than 170 bird species contribute to the bush soundtrack.

Wake with the birds at one of the park's 16 campgrounds, with most visitors opting to make the most of the facilities (including amenities for people with disabilities) at the Castle Rock and Kambuwal campgrounds near the visitor centre. The park's high elevation (approximately 1000m/3280ft above sea level) means it can get cold in winter, but with the access road passing several wineries, there's ample opportunity to grab a warming bottle of shiraz on your way in.

With a dozen walking trails in the park, it's worth planning to spend a good few days here – even just to laze around with the local eastern grey kangaroos.

🔵 THE PITCH

Camp amid the high-altitude splendour of this national park known for its immense granite boulders, springtime wildflowers and proximity to Stanthorpe-region wineries.

_____

When: year-round
Amenities: toilets, showers, barbecues and picnic tables at the main campgrounds
Best accessed: by car
Cost: $7.25
Contact: https://parks.des. qld.gov.au

07

WALLAMAN FALLS CAMPING AREA
GIRRINGUN NATIONAL PARK

08

It's a steep, winding drive up through Wet Tropics rainforest to reach Wallaman Falls (554m/1817ft above sea level), but it's well worth the detour off the Bruce Highway to feast your eyes on Stony Creek plunging a whopping 268m (879ft) off a sheer escarpment into the rainforest far below. After the strenuous 3.2km (2-mile) return hike to the base of the falls to really feel its power (and maybe take a restorative dip), Girringun National Park's only campground is an incredibly pleasant place to spend the night on the traditional lands of the Warrgamaygan people, whose spiritual ancestors dwell in the area.

Cradled by Stony Creek and shaded by open woodland, well-spaced campsites ring a grassy knoll dotted with picnic tables and wood-fired barbecues. A short pathway takes you through a tangle of upland rainforest to the creek, where you can also swim, or try your luck spotting an elusive platypus. Look out for vibrant crimson rosellas around camp, and if you packed a good torch, you can try spotlighting for brushtail possums and adorable sugar gliders after dark as a choir of rainforest frogs call to their mates, and bandicoots scurry around the undergrowth.

🔘 THE PITCH

Marvel at Australia's highest single-drop waterfall, then turn in for the night at the national park's only campground, where you might spot a musky-rat kangaroo as you pitch your tent.

When: May–Oct
Amenities: untreated water, toilets, showers, barbecues, picnic tables
Best accessed: by car
Cost: $7.25
Contact: https://parks.des.qld.gov.au

09

WILSON ISLAND
SOUTHERN GREAT BARRIER REEF

A vivid pop of green in the aquamarine shallows of the Southern Great Barrier Reef, tiny Wilson Island is home to one of Australia's coolest castaway experiences. Some 80km (50 miles) from the mainland, nine adults-only glamping tents line a crushed-shell beach, each naturally cocooned by pandanus and pisonia trees for privacy.

Stylish but small tent rooms encourage outdoor exploration: drifting in clouds of colourful fish, lazing in your hammock with a good book, or relaxing on your private deck with binoculars at hand for close-ups of seabirds that nest on the vegetated isles of Capricornia Cays National Park. That also includes nearby Heron Island, where you'll make a pit stop on the boat trip here from Gladstone – often a whole-day affair that only adds to the adventure.

Gourmet meals and drinks are included, and with no wi-fi or mobile/cell reception, plus only USB charging available, laptops are best left at home. While the island is closed to visitors during the peak bird breeding season, guests visiting from November to January can look forward to spotting marine turtles, which call the lagoon home year-round.

💰 THE PITCH

Watch marine turtles nest by moonlight just steps from your tent at this glamping getaway on a postcard-perfect Great Barrier Reef island.

When: Apr–Jan
Amenities: drinking water, restaurant, bathrooms, snorkelling equipment
Best accessed: by boat
Cost: $2590 for 2pp for 2 nights
Contact: www.wilsonisland.com

© WILSON ISLAND

NIGHTFALL CAMP
LAMINGTON NATIONAL PARK, SOUTHEAST QUEENSLAND

One of Australia's cleanest freshwater creeks meanders alongside Nightfall Camp, bouncing between smooth, sun-drenched boulders and flowing into serene natural swimming pools. It's just one of the many highlights of this glamping retreat at the gateway to the subtropical wilds of Lamington National Park.

Nestling under a dramatic escarpment, four studio-sized glamping tents hand-built by the owners immerse guests in an ancient landscape brimming with wildlife, with red-necked wallabies often seen munching on the verdant lawn beside the permaculture garden that drives the retreat's gourmet menu.

There's a two-night minimum stay, and when you arrive, it's easy to understand why. Backed by the peaceful sounds of the burbling creek, you'll feel so grounded by the serenity of the setting, you won't want to leave. Take your breakfast alfresco by the creek, then stretch your legs on local hikes, take an energising dip in the creek, or simply lounge on your private deck, watching cockatoos flap through bluebird skies. Darkness dials up a chorus of frogs, the soothing crackle of your wood-burning heater and the odd thump-thump of 'roos bounding through the bush.

😎 THE PITCH

Aussie bush meets ancient Gondwana rainforest at this luxurious, minimal-impact glamping retreat on the isolated southwestern edge of World Heritage-listed Lamington National Park.

When: year-round
Amenities: drinking water, private bathroom, restaurant, guest lounge with bar and wi-fi
Best accessed: by car
Cost: $1870 for 2pp for 2 nights
Contact: www.nightfall.com.au

© NIGHTFALL CAMP

10

THORSBORNE TRAIL CAMPING
HINCHINBROOK ISLAND NATIONAL PARK, CASSOWARY COAST

Separated from the mainland by a narrow channel, hulking Hinchinbrook Island is seriously wild. Protected since 1932, the 339 sq km (131 sq miles) rainforest-clad island known as Munamudanamy to the Bandjin and Girramay peoples looks much the same as it did when their ancestors gathered on its beaches to feast on shellfish you'll see in ancient middens dotting its rugged coast.

Tracing the island's eastern shoreline is the Thorsborne Trail. One of Australia's most famous multi-day hikes, the challenging 32km (20-mile) trail can typically be completed in four days, but there are seven bush campsites dotting the trail if you prefer to take it slow.

Hike to an ancient rhythm in this veritable Jurassic Park, where the island's mountainous backbone shields you from civilisation, and the surreal blues of the Coral Sea may make you curse the presence of crocs. With camping facilities limited to pit toilets and pack racks to keep your gear out of reach of opportunistic goannas, this isn't a trek for the faint of heart. But as you gaze out over the Great Barrier Reef Marine Park from the natural infinity pool at Zoe Falls, you'll be glad you made the sweaty scramble.

THE PITCH

Only 40 people at a time are allowed on the camping trail crossing this wild island between Cairns and Townsville, so book ahead and prep for an intimate experience.

When: Apr–Sep
Amenities: toilets at most campgrounds
Best accessed: by boat
Cost: $7.25
Contact: https://parks.des.qld.gov.au

11

© CORAL BRUNNER | SHUTTERSTOCK

© SARAH REID

ELIOT FALLS CAMPGROUND
APUDTHAMA NATIONAL PARK, CAPE YORK

Towards the pointy end of Cape York, Apudthama (formerly Jardine River) National Park is one of the most popular stops on a road trip to The Tip. Dominated by the Jardine River, which flows year-round, this vast and relatively flat sandstone landscape is threaded with crystal-clear streams and scenic waterfalls offering respite from long, dusty days on the road.

Of the park's five campgrounds, Eliot Falls is the standout, with three small waterfalls (Eliot Falls, Twin Falls and the Saucepan) all within a short stroll from the shady, red-dirt campground. Linked by Eliot Creek, the freshwater pools below the falls are all croc-free, making swimming in this remote oasis obligatory. Enjoy a yarn with fellow campers as you set up in one of the 31 campsites – every road tripper has a tale to tell by this point of their journey.

Eliot Falls is only 8km (5 miles) off the main road, but its location on the notorious high-clearance 4WD-only Old Telegraph Track means it's off-limits to conventional vehicles. Don't be in too much of a rush to get there, because you shouldn't miss the exquisite and refreshingly accessible Fruit Bat Falls, a broad waterfall with a large, crystalline pool accessed via the same main-road turn-off.

🅿 THE PITCH

Roll out of your tent to float in some of the most idyllic croc-free waterholes on the Cape York Peninsula, reached by one of the nation's most infamous 4WD tracks.

When: May–Oct
Amenities: toilets, picnic tables, firepits
Best accessed: by car
Cost: $7.25
Contact: https://parks.des.qld.gov.au

Indigenous Experiences

From eye-opening rainforest walks to gourmet sailing adventures, Queensland brims with new ways to experience the world's oldest living cultures.

Never before has there been such a rich array of options to connect with Australia's story through Indigenous tourism – especially in Queensland, the only state or territory where both Aboriginal and Torres Strait Islander cultures can be experienced on traditional lands.

Aboriginal and Torres Strait Islander Australia is made up of many different groups (known as Nations), each with their own culture, customs, language and laws passed down through generations. This means that no two Indigenous tourism experiences are the same, even when hosted in the same region, as each guide is a custodian of a unique ancestral heritage. Walking tours are a fantastic way to engage with culture in a short time with a minimal carbon footprint, but Indigenous tourism experiences across Australia now come in all shapes and sizes, from astronomy tours to art classes, high-tech drone shows to fishing lessons – each revealing unique insights into how Australia's First Peoples have lived in harmony with the natural world for millenniums.

CONNECTING WITH COUNTRY

Country is a term often used by Indigenous Australians to describe the lands and waterways to which they are connected. Country doesn't just refer to the physical land, but everything that inhabits it, from plants to animals to Dreaming (also known as Creation) spirits. Australia's Indigenous peoples have cultural responsibilities to care for their Country – an ancient system of sustainable environmental management

© CHRIS CHEN, MATT MUNRO, EWEN BELL | LONELY PLANET IMAGES

that continues today. Nothing deepens your own connection to Queensland and beyond like joining a Traditional Custodian on Country to learn about their relationship with the land. Two great places to search for and book Indigenous tours across the country include https://experience.welcometocountry.com and www.discoveraboriginalexperiences.com.

BEYOND INDIGENOUS TOURS

Seeking out Indigenous-owned galleries, museums, cafes, shops, accommodation and other businesses across Australia offers opportunities to both expand your understanding of Australia's Indigenous cultures and support the preservation and continuation of this rich cultural heritage.

Go crabbing on the coast (top) or meet an Aboriginal elder (above); an Indigenous Kuku Yalanji guide in the Daintree; traditional tools (inset)

FIVE TO TRY

Walkabout Cultural Adventures, Daintree
Immersive rainforest safari.
www.walkaboutadventures.com.au

Dreamtime Dive & Snorkel, Cairns
Great Barrier Reef with Indigenous guides.
www.dreamtimedive.com

Talaroo Hot Springs, Gulf Savannah
Connect with Ewamian Country on a guided tour.
www.talaroo.com.au

Taribelang Bunda Cultural Tours, Bundaberg
Time travel to sacred sites.
www.bundaculturaltours.com.au

Saltwater Eco Tours, Mooloolaba
Unique marine experiences.
www.saltwaterecotours.com.au

ROKO ISLAND GLAMPING
TORRES STRAIT

Turquoise waters swirl around the shallow coral reefs and tropical islands of the Torres Strait linking the Cape York Peninsula to Papua New Guinea. Just a 20-minute boat ride from the mainland village of Seisia, privately-owned Roko Island offers an incredible way to experience this watery world in rustic comfort. Here the Tchen Pan family operates one of the nation's most remote glamping stays, with four spacious tents lining a rocky shoreline patrolled by sharks and salties (saltwater crocodiles). While swimming is strictly off-limits, there are wild beaches and verdant mangroves to explore, or you can try your luck fishing off the jetty. Your host Jason can also arrange fishing charters as well as visits to nearby Thursday or Horn Islands to connect with Torres Strait Islander cultures and WWII heritage.

Served in a rustic outdoor dining area with sandy floors and superb views across the sparkling water, delicious three-course dinners include fresh-caught fish, from Spanish mackerel to prized coral trout. Order a drink from the beach bar to sip by a crackling fire in front of your tent, not that you need any extra warmth in the far reaches of Australia's tropical north, but it's fun all the same.

🍽 THE PITCH

It's an adventure just to get to Australia's northernmost glamping stay, where crocodiles patrol the tropical waters of a former pearl farm and fresh fish is always on the menu.

When: May–Oct
Amenities: drinking water, shared bathrooms, electricity, restaurant
Best accessed: by boat
Cost: $2400 for 2pp for 2 nights including meals and transfers
Contact: www.rokopearls. com.au

13

© SARAH REID

SPICERS SCENIC RIM TRAIL ECO HUTS
SCENIC RIM, SOUTHEAST QUEENSLAND

Tracing a rainforest-clad ridgeline in Main Range National Park, part of the UNESCO-listed Gondwana Rainforests, the Scenic Rim Trail opened in 2020 in a partnership between the QLD Parks & Wildlife Service and boutique hotel company Spicers Retreats. As part of the deal, the stunning trail is publicly accessible, with three basic walkers' camps with toilets located along the 47km (29-mile), one-way route, designed to be hiked from north to south over four days.

Requiring hikers to be completely self-sufficient, the independent option can be challenging. A sizeable step up in comfort (and price) is the guided Spicers Scenic Rim Trail, which includes two nights lodging in architect-designed 'eco camps' tucked off the main trail.

Entirely off-grid and meticulously designed to ensure a minimal environmental footprint, the camps feature minimalist cabin-style rooms, each with a floor-to-ceiling horizontal window that opens out fully into the rainforest. With just the fly screen closed at night, it connects you to the outdoors even more deeply than staying in a tent.

With sumptuous meals served in a communal dining room, and your luggage shuttled ahead to the next camp each day, the extra touches on the guided experience compensate for the very real possibility of encountering a leech or three on this lush trail.

14

© PAUL HARRIS

THE PITCH

Taking you deep into one of the world's most significant rainforests, the guided version of Queensland's newest multi-day hike includes memorable nights in state-of-the-art eco-cabins.

When: Apr–Nov
Amenities: drinking water, shared bathrooms, electricity, wi-fi
Best accessed: on foot
Cost: camping from $7.25; guided 4-night walk $3999 pp
Contact: https://parks. des.qld.gov.au; www. scenicrimtrail.com

15

COBBOLD CAMPGROUND
COBBOLD GORGE, GULF SAVANNAH

Queensland's youngest gorge is an oasis in the heart of Gulf Savannah country, formed by water funnelling through ancient cracks in the sandstone landscape over 1700 million years. Unknown to generations of station owners and workers until the Terry family stumbled across it on their property in 1992, the narrow gorge on the traditional lands of the Ewamian people has evolved into one of the state's most popular outback attractions, with tours taking visitors deep into the geological wonder by boat and stand-up paddleboard. There's even a glass bridge spanning a section of the gorge, 19m (62ft) above the cool sapphire water, home to a handful of freshwater crocs.

The property's pleasant, red-dirt campground offers some of the best camping facilities for miles. Choose from powered and unpowered sites suitable for tents, caravans and even big rigs; as well as rooms and cabins, including a wheelchair-accessible option. Pitch your tent to the sounds of parrots, finches and honeyeaters chattering in the registered nature refuge, and keep an eye out for kangaroos, wallaroos, echidnas and goannas on a hike on Cobbold Gorge's four walking trails.

🟡 THE PITCH

Spend an invigorating day touring an unusual outback gorge before flopping into a tent at this well-equipped campground — perhaps via a dip in the pool.

When: Apr–Oct
Amenities: drinking water, camp kitchen, swimming pool, laundry, bar/restaurant, barbecues, wi-fi
Best accessed: by car
Cost: from $18
Contact: www.cobboldgorge.com.au

MUNGA-THIRRI NATIONAL PARK
SIMPSON DESERT

16

Spectacular sand dunes in a palette of warm hues stretch to oblivion in Munga-Thirri National Park. Spanning a whopping one million hectares, this vast desert landscape encompasses but a fraction of the iconic Simpson Desert, which covers more than 17 million hectares of Central Australia.

For seriously well-prepared travellers with sand-driving experience and mates to travel in convoy with as recommended by rangers for safety, this is your chance to experience the essence of off-grid camping. Drop into the historic Birdsville Hotel for a pint before heading into the otherworldly park. Snap a photo atop Big Red – an immense sand dune rising 30m (98ft) above the arid plains – before venturing off into the desert wilderness to find the perfect place to spend the night, with bush camping allowed within 100km (62 miles) of the QAA Line, a 4WD track connecting Birdsville to the Northern Territory.

Spend the cooler late-afternoon hours tracking a charismatic thorny devil, perentie (desert goanna) or tiny spinifex hopping mouse by footprints in the red sand, or sink into a camp chair and watch white-winged fairy-wren flitting between clumps of sandhill canegrass. With little need for a tent out here, simply pack a swag to roll out beneath the celestial canopy.

🔆 THE PITCH

Experience the ultimate sense of freedom on a 4WD camping adventure deep in the ancient longitudinal desert dunes of Australia's largest national park.

When: Apr–Sep
Amenities: none
Best accessed: by car (4WD)
Cost: $7.25
Contact: https://parks.des. qld.gov.au

WHITSUNDAY ISLANDS NATIONAL PARK
WHITSUNDAYS

Rising up from the cyan waters off Airlie Beach like a cluster of emerald jewels are the 74 Whitsunday Islands. Blanketed in dense rainforest, laced with walking tracks and fringed by powder-white beaches, most of the islands form part of the Whitsunday Islands National Park, and rustic beach camping is allowed at 11 sites across three islands within its boundaries.

And you don't even need your own boat to reach them, with a number of Airlie Beach operators offering transfers.

Roll out of your tent on Crayfish Beach on the eastern side of Hook Island to snorkel in Mackerel Bay, which has some of the best coral cover in the Whitsundays. Or set up camp at peaceful Curlew Beach, also on Hook Island, the closest camping area to the Ngaro Cultural Site with its precious rock art and middens. Or enjoy the pillow-soft sands and aquamarine waters of Whitsunday Island's world-famous Whitehaven Beach all to yourself (and your fellow campers) after the day trippers have departed. The winter and spring months tend to offer the best weather, although you should be mindful that stinger season begins in October. With no water, food or barbecues available at the beach campsites, you'll need to be well prepared and completely self-sufficient.

17

🔘 THE PITCH

Embrace the castaway feeling of a camping adventure to an uninhabited island in the Whitsundays, where every bush campground has a unique selling point.

When: May–Oct
Amenities: toilets, picnic tables
Best accessed: by boat
Cost: $7.25
Contact: https://parks.des.qld.gov.au

TINY WILD HOUSES
SOUTHEAST QUEENSLAND

When unprecedented bushfires swept through Lamington National Park in September 2019, the worst fears of Binna Burra Lodge Chairman Steve Noakes were realised when the historic main lodge of the iconic resort surrounded by the national park was lost to the blaze.

As the park's Gondwana Rainforests continue to recover, Binna Burra Lodge has also bounced back with new accommodation options including its Tiny Wild Houses, a string of compact, eco-sensitive cabins perched on the Bellbird clifftop, some 800m (2625ft) above sea level. Wake naturally to the morning light streaming in through the floor-to-ceiling windows before pulling on your hiking boots: more than half of the national park's 21 trails are accessible from your doorstep. As the colours of sunset fade and the kookaburras have their last laugh of the day, the lights of the Gold Coast can be seen twinkling on the horizon.

Binna Burra Lodge also offers camping, glamping and modern lodge rooms. There are 10 remote, hike-in bush camps in the surrounding national park, as well as camping, glamping and rooms at O'Reilly's Rainforest Retreat, a historic lodge located in the Green Mountains section of the park.

🔖 THE PITCH

Risen from the ashes of Australia's 2019–20 bushfires is this suite of tiny stays at Binna Burra Lodge, offering a new way to experience the splendour of Lamington National Park.

***

When: year-round
Amenities: drinking water, bathroom, kitchenette, wi-fi
Best accessed: by car
Cost: from $377 p/cabin
Contact: www.binnaburralodge.com.au

18

© TINY WILD HOUSES

GHEEBULUM KUNUNGAI (MORETON ISLAND) NATIONAL PARK
MULGUMPIN (MORETON ISLAND)

The bright lights of Brisbane fade to a dull glow across the shallow waters of Moreton Bay as darkness falls on Mulgumpin (Moreton Island), and the haunting cries of the critically endangered eastern curlew ring out among the melaleucas. This vast and largely undisturbed sand island on Quandamooka Country is best known for its Tangalooma wrecks, a string of scuttled boats perfect for snorkelling around. The wrecks can be visited on a day trip from the city, but it's much more fun overnighting at one of the 10 campgrounds scattered through the protected recreation area that covers most of the island. The Micat ferry drops you next to the wrecks, with a great bush campsite just behind the beach. For more seclusion, load your 4WD onto the ferry and follow sandy tracks to more remote campgrounds on the island's west, north, and more exposed east coasts. A series of walking trails threads through this ecologically and culturally important landscape, leading to attractions including the serene Blue Lagoon, a freshwater oasis set amid flowering heathland, and Mt Tempest (280m/919ft), the highest sand dune on the island.

Don't forget to pack a portable cooker, with open fires prohibited on the island year-round.

19

🔵 THE PITCH

The only thing better than a day trip to the world's third-largest sand island is an overnight stay in one of its bush campgrounds tucked behind its snow-white beaches.

When: year-round
Amenities: main sites have untreated water, toilets, showers and rubbish bins
Best accessed: by ferry
Cost: $7.25
Contact: https://parks.des.qld.gov.au

20

NOOSA RIVER CANOE CAMPING

COOLOOLA RECREATION AREA, SUNSHINE COAST

Queensland boasts two UNESCO-listed biosphere reserves, and they both meet in the Sunshine Coast holiday hotspot of Noosa. The Noosa Biosphere Reserve alone contains 61 distinct ecosystems, and this is where you'll launch a kayak or canoe (both available for rent locally) and paddle along the turquoise curves of the Noosa River for up to 40km (25 miles) from Boreen Point to make camp at one of nine national-park campsites set back from the river's sandy banks. The tea-tree-stained upper reaches of the river form part of the Great Sandy Biosphere, also known for its ecological diversity and prolific birdlife. Lap up the lush seclusion, where binoculars come in handy for closer observation of the many bird species that may visit your private campsite (each with room for four people) among the banksias and twisted swamp paperbarks. Motorised boats are not allowed to travel upriver past campsite three, leaving a chorus of frogs to set the tone at dusk.

Upper Noosa River campsites 1–3 have toilets; beyond this point you'll need a portable toilet or a poop kit to bury your business responsibly. The 102km (63-mile) Cooloola Great Walk winds past campsites 3–5; further on, the river is the only trail in the wilderness.

🛶 THE PITCH

Load up a canoe or kayak for a paddling and camping adventure up the Sunshine Coast's Noosa River, where two biosphere reserves meet in an explosion of birdsong.

When: year-round
Amenities: varies
Best accessed: by kayak or canoe
Cost: $7.25
Contact: https://parks.des.qld.gov.au

BOWRA WILDLIFE SANCTUARY
SOUTHERN QUEENSLAND OUTBACK

The Australian Wildlife Conservancy is dedicated to the conservation of all native animal species and the habitats in which they live, with visitors welcomed at 11 of its sanctuaries – several of which allow camping. A twitcher's nirvana, Bowra Wildlife Sanctuary is one of them.

The dusty town of Cunnamulla sits at the gateway to this outback wetland, where more than 200 bird species are the premier attraction. The rare grey falcon has been known to breed at Bowra, but for many visitors the parrots and fairy wrens that flock here are a highlight, with their colourful plumages contrasting beautifully against the semi-arid landscape. An array of reptiles, 'roos, and plants also thrives here.

Join staff for the Daily Bird Call survey, with bird observation forms available. Or head deeper into the sanctuary on a self-guided 4WD safari before settling into the rustic red-dirt campground as the park's birds head to their roosting spots, and bats take over the skies. Benches beside a nearby lagoon make a great spot to enjoy a birdwatching breakfast.

With all proceeds from Bowra's overnight guests reinvested into the campground and the conservation of local wildlife, think of it as positive-impact camping.

● THE PITCH

Play a role in the conservation of native birds and wildlife with an overnight visit to this Australian Wildlife Conservancy-run outback sanctuary.

When: May–mid-Oct
Amenities: untreated water, toilets, showers, limited electricity
Best accessed: by car
Cost: $15–25, 2-night minimum
Contact: www.australianwildlife.org

© WAYNE LAWLER

22

BABINDA BOULDERS FREE CAMPING AREA
SOUTH OF CAIRNS

Gin-clear water pools at a bend of Babinda Creek at Babinda Boulders, forming a natural swimming hole so beautiful you'll want to jump straight in. With neighbouring Wooroonooran National Park providing a serene rainforest setting, this popular recreation area sits right across the (sealed) road from one of Australia's best free camping areas, where eight spacious sites fan out from a tight loop road encircling a toilet block. With no booking system, you'll need to get here early to snag a spot in the campground, and if you do, you can stay for up to three days.

In the well-maintained day-use area, both the toilets and the swimming hole have all-abilities access. Less suitable for wheelchairs is the 1.3km (0.8-mile) return walk downstream to two viewing platforms where the creek cascades into washpools known as the Devil's Pool. Find your own private swimming hole on a walk upstream from the day-use area along Wooroonooran National Park's Goldfield Trail, which features a peaceful hike-in campsite on the Mulgrave River.

🏕 THE PITCH

Looking for an easy nature-based escape? Just steps from a blockbuster freshwater swimming hole, one of Tropical North Queensland's most accessible campgrounds won't cost you a cent.

When: May–Oct
Amenities: untreated water, toilets, showers, barbecues
Best accessed: by car
Cost: free
Contact: www.cairns.qld.gov.au

CAPE HILLSBOROUGH NATURE TOURIST PARK
MACKAY REGION

Rugged, rainforest-cloaked hills plunge to sandy beaches lapped by turquoise water in Cape Hillsborough National Park, just north of Mackay. Bookended by this coastal wilderness, Cape Hillsborough Nature Tourist Park offers a front-row seat to the sunrise spectacle of agile wallabies foraging on the adjacent beach. With the macropods now fed special 'roo food by the local tourism board, visitors looking for a more authentic immersion in the traditional lands of the Yuibera people would be wise to hit the national park's four short but ultra-scenic trails. From the edge of the tourist park, the steep and rocky Andrew's Point Track (2.8km/1.7-mile circuit) is the pick of the bunch, climbing up through moist vine forest to the top of a headland for incredible vistas towards the Cumberland Islands. Don't miss the Turtle Lookout, where you're bound to spot at least a few turtles swimming in the cyan water far below.

Right next to the beach, the unpowered camping area almost makes you forget you're staying in a large, commercial holiday park. Facing directly east, its sunrise vistas are all-time.

If you don't mind driving to the trailheads, stay at the more peaceful Smalleys Beach campground on the national park's northern shores.

23

🗨 THE PITCH

Cape Hillsborough is best known for its beach-loving wallabies, but the underrated highlight of this coastal campground is the excellent walking trails that lace the national park surrounding it.

When: year-round
Amenities: drinking water, toilets, showers, barbecues, picnic tables, laundry, cafe, rubbish bins
Best accessed: by car
Cost: from $33
Contact: www.cape hillsboroughresort.com.au

RANGELANDS OUTBACK CAMP
WINTON

When you need a break from pitching a tent, this swish glamping stay just outside Winton offers a stylish slice of outback Queensland indulgence. With room for just 12 guests between six spacious, solar-powered tents (complete with air-con and rustic-chic ensuites), a stay at Rangelands Outback Camp is an intimate experience, with gourmet meals (included in the price) served on a fabulous deck with an undisturbed view of Winton's stellar sunsets, followed by star-spangled night skies.

So superb is the night sky here that it's the home of the Jump-Up, Australia's first Dark Sky Sanctuary, certified in 2019 and just a 30-minute drive from Rangelands Camp. Located on a rugged mesa plateau, the Jump-Up doubles as the home of the excellent Australian Age of Dinosaurs Museum. At night you can visit the museum's Star Gallery for free, or get better acquainted with the cosmos on its Gondwana Stars Observatory tour, which includes telescope viewing opportunities. But you can also take a dazzling celestial journey without even leaving Rangelands – the night sky sparkles beautifully here, too.

🌀 THE PITCH

Gaze into Queensland's darkest skies at an outback-luxe glamping stay that makes the perfect base for exploring the Winton region's famed Australian Dinosaur Trail.

Best time: mid-Mar—Oct
Amenities: drinking water, restaurant, private bathroom, electricity
Best accessed: by car
Cost: from $690
Contact: https://rangelandscamp.com

24

SOUTH AUSTRALIA

Seals, whales, emus, kangaroos and koalas are just some of the locals you'll meet on immersive stays spanning South Australia's vast outback to its wild coasts.

Best time: Apr–Sep (outback, Nullarbor, Flinders Ranges); year-round (coastal regions)
Best national parks for camping: Coffin Bay National Park, Deep Creek National Park, Dhilba Guuranda-Innes National Park, Gawler Ranges National Park, Mount Remarkable National Park
Best camping trails: Heysen Trail, Kangaroo Island Wilderness Trail
National parks pass required: Desert Parks Pass for outback parks; some other parks require a vehicle pass
Useful contacts: www.south australia.com; www.parks.sa.gov.au

A state of captivating contrasts, South Australia's arid outback meets paradisiacal beaches, verdant wineries surround ancient limestone caves and rugged ranges rise up from sweeping savannas. Add oodles of wildlife and some of the world's clearest skies – the state boasts a dark sky sanctuary *and* a dark sky reserve – and it makes for sublime camping, with campgrounds in more than 40 of its national parks.

South Australia's short but hot summers and frosty winters making camping adventures more popular in autumn and spring (hint: book ahead). Yet winter is great for walking and whale-watching, while beaches are at their most inviting at the height of summer.

FREE CAMPING

South Australia has a good range of free camping sites, typically in designated areas and including plenty along the Nullarbor. Always check local council rules before you get cosy – a fee was recently introduced at Perlubie Beach on the Eyre Peninsula, for example, with numbers also capped to manage its rising popularity.

SUPPLIES

Adelaide/Tarndanya is the best place to stock up, with all the big brands represented, from BCF to Snowys Outdoors to 4WD Supacentre. Operating since 1909 under various names, independent outfitter Exurbia offers a huge range of gear at its Norwood store (www.exurbia.com.au). Remember to gather or purchase firewood before you enter national parks, where collection is prohibited.

SAFETY

Hot days can quickly turn into bone-chilling nights in South Australia, making sun protection and cold-weather gear essential year-round. While there are no crocs to contend with, it's generally wise to keep your distance from kangaroos and emus, which can be aggressive if threatened. And be sure to stay abreast of bushfire warnings and seasonal fire bans.

BEST REGIONS

Kangaroo Island

The isolation and remoteness of 'KI' for thousands of years produced a special environment where rare and endemic wildlife continues to thrive. And there is no shortage of places to stay surrounded by its natural beauty.

Vines in McLaren Vale, one of South
Australia's wine regions (left), 'roos (below)

Flinders Ranges

The 600-million-year-old
peaks and rocky gorges of
the Flinders Ranges in outback
South Australia form some of
the nation's most dramatic
landscapes. Rich in Aboriginal
and pastoral history, the region
is also home to a vast array of
wildlife – and some memorable
places to sleep under the stars.

Eyre Peninsula

Fringed by more than
2000km/1243 miles of coastline,
this enormous, triangular
peninsula west of Adelaide offers
South Australia's best beach
camping. But it's also got outback
options in the Gawler Ranges to
the north.

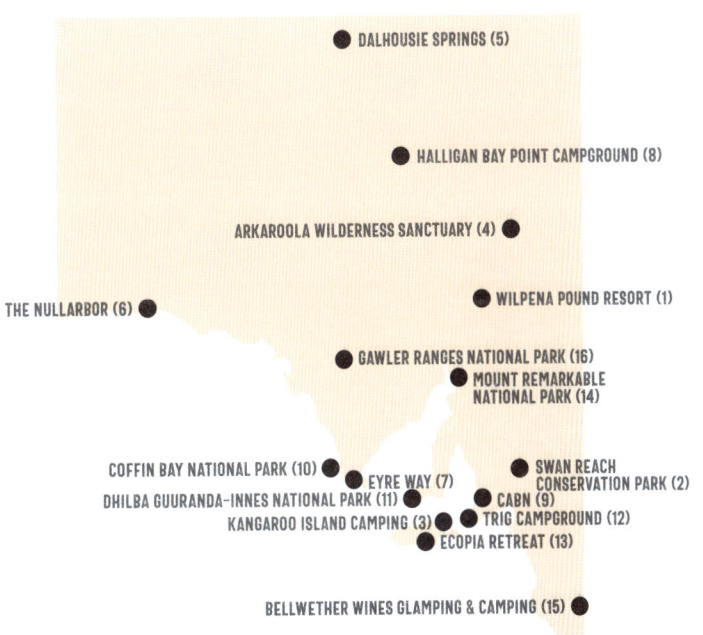

DALHOUSIE SPRINGS (5)

HALLIGAN BAY POINT CAMPGROUND (8)

ARKAROOLA WILDERNESS SANCTUARY (4)

THE NULLARBOR (6)

WILPENA POUND RESORT (1)

GAWLER RANGES NATIONAL PARK (16)

MOUNT REMARKABLE
NATIONAL PARK (14)

COFFIN BAY NATIONAL PARK (10)

EYRE WAY (7)

SWAN REACH
CONSERVATION PARK (2)

DHILBA GUURANDA-INNES NATIONAL PARK (11)

CABN (9)

KANGAROO ISLAND CAMPING (3)

TRIG CAMPGROUND (12)

ECOPIA RETREAT (13)

BELLWETHER WINES GLAMPING & CAMPING (15)

WILPENA POUND RESORT
IKARA–FLINDERS RANGES NATIONAL PARK

Some 400km/250 miles north of Adelaide, on the edge of South Australia's outback, rises Wilpena Pound – a natural ring of mountains resembling a giant crater. Home to rugged mountain vistas and 500-million-year-old fossils (believed to be remnants of prehistoric sea life), Wilpena Pound remains an enormously significant place to Adnyamathanha/Yura peoples, who have lived in the surrounding Flinders Ranges for tens of thousands of years. Since 2012, they have owned Wilpena Pound Resort – the only place to stay inside the national park protecting this special place.

Choose from glamping safari tents, powered and unpowered sites and rooms, with access to the resort's restaurant, bar, pool and store. Experience a traditional Welcome to Country, hosted every evening, then join an Adnyamathanha guide on a tour of sacred sites in the park including 40,000-year-old rock paintings at Arkaroo Rock, and rock engravings estimated to be even older. Learn about the intertwining histories of Aboriginal and European cultures in this bush landscape, where failed farms now form part of the national park, a refuge once again for wallabies, kangaroos, emus and other wildlife.

● THE PITCH

Backdropped by a geological wonder, the only accommodation within Ikara-Flinders Ranges National Park offers myriad opportunities to connect with culture and the surrounding landscape.

When: Apr–Oct
Amenities: drinking water, toilets, showers, electricity, restaurant, swimming pool, store, waste facilities, laundry, wi-fi
Best accessed: by car
Cost: camping from $38
Contact: www.wilpenapound.com.au

© WILPENA POUND RESORT

SWAN REACH CONSERVATION PARK
RIVER MURRAY INTERNATIONAL DARK SKY RESERVE, MID MURRAY

02

Night sky darkness is measured at a level of between 0-22, with Sky Quality Meter (SQM) readings closest to 22 among the darkest places on the planet. Just a 90-minute drive from Adelaide, the 3200 sq km (1235 sq miles) River Murray International Dark Sky Reserve (RMIDSR) consistently measures about 21.8, with the Mt Lofty Ranges creating a natural barrier from the urban lighting in South Australia's capital, and the dry climate and low humidity in the area creating the ideal conditions for optimum stargazing year-round, particularly during winter.

At the core of the RMIDSR, which includes a number of townships, nature reserves, private farms and an 80km (50-mile) stretch of the Murray River, is the 20 sq km (7.7 sq miles) Swan Reach Conservation Park. Here SQM readings have registered an astronomically dark 21.95.

Bring everything you'll need to spend a night camping under the Milky Way at any spot in this semi-arid mallee landscape that takes your fancy. The public entrance to the park is off the Stott Highway, around 15km (9.3 miles) west of Swan Reach, and its dirt roads are navigable by 2WD. If you're lucky, you might stumble across a southern hairy-nosed wombat at your campsite – these elusive mammals were the main reason for the proclamation of this park in 1970.

🧭 THE PITCH

Free camp at the core of Australia's first internationally accredited dark sky reserve and prepare to be dazzled by the nightly show that plays out in its crystal-clear skies.

When: year-round
Amenities: none
Best accessed: by car
Cost: free
Contact: www.rivermurraydarkskyreserve.org; www.parks.sa.gov.au

KANGAROO ISLAND CAMPING
KANGAROO ISLAND

03

Sea lions cavort on snow-white beaches, koalas doze in twisted gum trees, and echidnas waddle, well, just about everywhere on Kangaroo Island. And these are only a few of the wildlife sightings you can expect to chalk up on a camping trip to this 4405 sq km (1700 sq miles) island, also known for its artisan produce.

Surrounded by the red cliffs of Flinders Chase National Park, the remote West Bay Campground is a wonderfully secluded place to pitch a tent amid roaming Rosenberg's goannas, with just eight sheltered campsites, toilets and a picnic area. Decimated by Australia's 2019-20 bushfires, the campsites on the Kangaroo Island Wilderness Trail that weaves through this epic coastal wilderness are slated to reopen in 2024. Also rebuilt after the fires are two charming heritage-listed cottages in the heart of the national park offering cosy bases for exploration.

Beyond the island's small handful of national park campgrounds, Kangaroo Island Council operates seven grounds with a range of facilities in similarly scenic locations, chief among them Vivonne Bay with its dreamy cyan water. These can't be booked ahead, so arriving early is key.

You can fly to Kangaroo Island and rent a car, but driving onto the ferry at Cape Jervis, less than two hours' drive south of Adelaide, only adds to the adventure.

© LEV KROPOTOV | SHUTTERSTOCK

🔶 🧭 THE PITCH

The dramatic scenery of Kangaroo Island surrounds a mix of privately-owned, council-run and national park camping areas where you can throw up a tent or even stay in a historic hut.

When: year-round
Best time: May–Oct
Amenities: varies
Best accessed: by ferry, then car
Cost: varies
Contact: www.parks.sa.gov. au; www.kangarooisland. sa.gov.au

04

ARKAROOLA WILDERNESS SANCTUARY
NORTHERN FLINDERS RANGES

Tucked up in the northern reaches of the rugged Flinders Ranges, Arkaroola Wilderness Sanctuary has operated with an emphasis on science, education and conservation for more than half a century. In 2023, this 610 sq km (235 sq miles) property, which offers powered and unpowered camping as well as bunkhouse and cottage-style accommodation, became Australia's second internationally certified dark sky sanctuary.

A dark sky sanctuary is typically situated in a very remote location; some 600km (373 miles) north of Adelaide, Arkaroola ticked the box. Here, campers can choose from a range of astronomy activities including an observatory tour and a stargazing experience. Or stump for the Ridgetop Sleepout – a hosted night under the stars, camping on one of only four sleepout decks positioned on top of a spectacular ridge. By day, there's bushwalking, 4WD trails to explore, wildlife to spot (Arkaroola is a particularly good place to see endangered yellow-footed rock wallabies), a pool to cool off in, and a restaurant and bar to tempt you away from the camp kitchen.

🔵 THE PITCH

On a remote private property, some 100km (62 miles) from the closest town, Australia's newest dark sky sanctuary offers astronomy experiences including a ridgetop sleepout under the stars.

When: year-round
Amenities: drinking water, toilets, showers, barbecues, swimming pool, waste facilities, electricity, wi-fi, laundry, EV charger, fuel, shop
Best accessed: by car
Cost: camping from $35
Contact: www.arkaroola.com.au

DALHOUSIE SPRINGS
WITJIRA NATIONAL PARK

On the western edge of the Simpson Desert, over 1200km (746 miles) north of Adelaide, the World Heritage-listed Dalhousie Springs are part of a chain of more than 120 mound springs extending along the outer rim of the Great Artesian Basin. Used by Aboriginal peoples for millennia, the ancient artesian water rises up through cracks and fissures in the subterranean strata, filling natural pools perfect for a dip in the 37°C water. After the long, dusty drive out here on the corrugated 4WD-only access road, it's a magical feeling to slip into the spring at sunset, when the wide, open sky erupts into colour. If you're visiting the park a few weeks after a soaking rain, you might also be lucky enough to see this desert landscape burst into bloom.

Just a short walk from the Main Spring is Witjira National Park's primary campground, with unallocated spaces for up to 50 vehicles. Expect company – it might be in the middle of nowhere, but this is one of Australia's most popular outback campgrounds. And it doesn't only attract people – you'll want plenty of mosquito repellent out here. If you're after more solitude (and fewer mozzies), head to the park's other campground, 3 O'Clock Creek, just 12km (7.5 miles) away.

05

THE PITCH

Miles from anywhere, this popular outback campground offers doorstep access to a UNESCO-listed hot spring, tailor-made for a soothing soak after a long drive.

When: Apr–Sep
Amenities: toilets, hot showers
Best accessed: by car (4WD)
Cost: $25.50
Contact: www.parks.sa.gov.au

© TED MEAD | GETTY IMAGES

THE NULLARBOR
GREAT AUSTRALIAN BIGHT

Gain a whole new appreciation for wide open spaces as you drive across the Nullarbor Plain on the Eyre Highway linking South Australia's Eyre Peninsula with the goldfields of Western Australia. Much of this wide, flat, semi-arid karst landscape – twice the size of South Korea – is protected by South Australia's Nullarbor National Park, Wilderness Protection Area and Regional Reserve. And it's possible to free camp in numerous signed locations throughout the protected areas, and outside them. You'll need a 4WD to follow the maze of tracks leading off the highway to camp beside the Nullarbor's soaring cliffs fringing the Great Australian Bight. And the cooler months are ideal for it, when southern right whales gather in huge numbers to breed, particularly at the Head of Bight, which has toilets and tank water, and a free basic campground nearby. You might chalk up a few whale sightings from your tent at some spots along the coast, but don't pitch up too close to the edge, as the cliffs are constantly eroding.

With only a couple of petrol stations and very limited facilities along the Eyre Highway – the only sealed road in the Nullarbor – this is an adventure that calls for a high degree of preparation.

THE PITCH

Find beauty in the isolation and stark surrounds of the vast Nullarbor Plain, where you can free camp beside soaring cliffs high above breeding whales.

When: mid-May–Oct
Amenities: none
Best accessed: by car
Cost: free
Contact:
www.parks.sa.gov.au

06

© ELECTRA | SHUTTERSTOCK

EYRE WAY
SLEAFORD, EYRE PENINSULA

First, in 2021, came Yambara, an architecturally designed, off-grid tiny abode on a secluded stretch of the Eyre Peninsula between Sleaford Bay and the dramatic cliffs of the Whalers Way. Then, in 2022, came its sibling Maldhi, located on the same private farm, just a 30-minute drive from Port Lincoln. While Yambara is light and bright with sunset tones, Maldhi is furnished in earthier olive hues. But both have the same layout, offering a front-row seat to the natural beauty of the Eyre Peninsula, and a private oasis to relax and recharge in supreme comfort, with a double shower and well-appointed kitchen at your disposal. Both cabins have a double bed and an additional bed in a mezzanine space ideal for kids, as well as a sun-drenched deck, perfect for watching whales cruise across Sleaford Bay during the winter months, and the odd 'roo hop by your cabin. There's also private-beach access for bracing swims as white-bellied sea eagles ride the breezes above.

As darkness descends it can be difficult to tell where the Southern Ocean ends and the sky begins as lights from local fishing boats twinkle on the water.

🟠 THE PITCH

Take in the raw beauty of the Eyre Peninsula while staying warm and cosy in this twin set of tiny abodes nestled on a remote stretch of private coastal farmland.

When: year-round
Amenities: drinking water, kitchen, electricity, wi-fi, firepit, barbecue
Best accessed: by car
Cost: from $1100, 2-night minimum
Contact: www.eyreway.com

07

© AMY ROWSELL PHOTOGRAPHY

HALLIGAN BAY POINT CAMPGROUND
KATI THANDA–LAKE EYRE NATIONAL PARK

Only 4WDs are allowed in Kati Thanda-Lake Eyre National Park – this is a serious adventure. Reached via the Oodnadatta Track, followed by another similarly bone-rattling 60km (37 miles) across Anna Creek Station, the park's only campground parks you right in front of shimmering Kati Thanda-Lake Eyre. Lying at Australia's lowest point, 15m (49ft) below sea level, this vast salt lake – its basin spanning nearly one sixth of the continent – is an important place for Aranbana people. Its Aboriginal name, Kati Thanda, describes how the lake was formed by spreading the skin of a kangaroo over the ground.

Halligan Bay Point is set at the tip of a small peninsula surrounded by the lake. After rare heavy rainfall, waterbirds descend in their thousands. As water evaporates, the lake's salinity increases, turning a photogenic strawberry-milk hue due to the bacteria that blooms in its shallow waters.

Unallocated camping is available for 20 vehicles. Bring everything with you – the lake water is too salty to drink, and swimming can be a stinging experience.

🗨 THE PITCH

Known to teem with birds after rain, and turn pink in drier conditions, Australia's largest lake offers a surreal bush camping experience on its salty shores.

When: Apr–Sep
Amenities: toilets, picnic tables
Best accessed: by car (4WD)
Cost: $21.50
Contact details: www.parks.sa.gov.au

Wine tasting

Nothing elevates a camp meal quite like a glass of local wine. And in South Australia there is no shortage of cellar doors (more than 350 and counting) to pick up a bottle for your adventure ahead.

Home to nearly half of Australia's vineyards and producing around 80 per cent of the nation's premium wine, South Australia is a wine-lovers' nirvana. Best known for its full-bodied reds, the state produces a wide variety of vintages across 18 wine regions, from the artisanal sparkling wines of the Adelaide Hills to crisp, citrus-laced rieslings that are synonymous with the Clare Valley.

Get a taste for South Australian wine at the National Wine Centre of Australia (www.nationalwinecentre.com.au) in Adelaide, then set out to visit your favourites at the source.

WINE TRAILS

Whether you prefer to discover wineries on foot, by bike or car, there are plenty of trails to choose from. Spend a fabulous few hours visiting the six wineries on the 5km (3.1-mile) Coonawarra Walking Trail on the Limestone Coast (https://coonawarrawalkingtrail.com.au), jump on a bike to explore the Clare Valley's 33km (21-mile) Riesling Trail (www.clarevalley.com.au), or pair fine wine with fabulous food on a multi-day Epicurean Way road trip (https://southaustralia.com/roadtrips).

SUSTAINABLE WINEMAKING

Just a 45-minute drive south of Adelaide, McLaren Vale is South Australia's oldest wine region, but this area, famed for its bold reds, is also one of Australia's most progressive and environmentally sustainable wine communities. Learn more at the likes of biodynamic winery Gemtree Wines (https://gemtreewines.com). The Adelaide Hills is another region known for its forward-thinking

viticulture, while over on the Fleurieu Peninsula, Hither & Yon (https://hitherandyon.com.au) is the state's first carbon-neutral certified wine brand. Cheers to that.

WHEN TO GO

With most South Australian cellar doors open year-round, it's always a good time to go winery hopping. But for foodies and wine lovers, autumn in wine country is an ideal season to experience end-of-harvest celebrations such as the Barossa Vintage Festival (April),

Tasting Australia in Adelaide (May) and the Clare Valley Gourmet Weekend (May). The vineyards are at their most photogenic in summer with vines laden with ripening fruit, while the cooler months are perfect for cosying up beside the cracking fireplaces found in many tasting rooms.

Vines in Clare Valley (top) barrels in the Barossa Valley at Seppeltsfield (above); wine tasting (left) and growing grapes in the Barossa (inset)

FIVE TO TRY

d'Arenberg Cube, McLaren Vale
Futuristic cellar door that stimulates the senses.
www.darenberg.com.au

Unico Zelo & Applewood Distillery, Adelaide Hills
Fun, wines plus craft gins.
www.applewooddistillery.com.au

Seppeltsfield, Barossa Valley
A Barossa mainstay with a range of tasting options.
https://seppeltsfield.com.au

Skillogalee Estate, Clare Valley
Tastings or long lunches in an 1851 cottage.
www.skillogalee.com.au

The Islander Estate Vineyard, Kangaroo Island
Relaxed winery in a remote, emerging wine region.
www.iev.com.au

CABN
VARIOUS LOCATIONS

Born in South Australia and now expanding across the country, CABN's self-contained off-grid tiny stays are designed to tread ultra-gently on the surrounding environment – all are powered by the sun, built to conserve water and furnished with eco-friendly amenities.

Generally located within a two-hour drive from a major city, they come in three types: tiny homes on wheels (CABN), glamping tents with outdoor bathtubs (CANVS), and larger, more luxurious cabins complete with private sauna (CABN X). There's no wi-fi in any of them, but founder and CEO Michael

Lampress guarantees a better connection – not with your favourite social media but with nature and yourself. Some properties are also dog-friendly.

Wake up to the vineyards of the Barossa Valley bathed in golden light, cosy up surrounded by towering pines in the Kuitpo Forest, or gaze over the Southern Ocean from your perch on Kangaroo

Island – blissfully removed from real-world distractions.

There's usually a two-night minimum stay, but single-night bookings are sometimes available within 30 days of the available date. A remote service fee is applied to every booking, and if you do need to connect, most properties receive mobile-phone reception.

🌱💲 THE PITCH

If you go down to the woods today – you may stumble across one of these minimalist-chic cabins by South Australia's original off-grid tiny stay company.

When: year-round
Amenities: drinking water, electricity, private bathroom, kitchenette, firepit, barbecue
Best accessed: by car
Cost: from $229
Contact: https://cabn.life

© CABN

COFFIN BAY NATIONAL PARK
EYRE PENINSULA

South Australia's Eyre Peninsula is blessed with an excellent range of places to camp amid wandering emus and snuffling echidnas. But if you really want to get away from it all, the windswept coastal wilderness of Coffin Bay National Park, at the southern tip of the peninsula, is the place to pitch a tent (or tow your camper trailer). There are seven campsites in the national park, all with their own draws. If you're visiting in a 2WD, Yangie Bay Campground makes choosing easy, as it's your only option. Located beside a protected inlet perfect for paddleboarding and kayaking, it also has firepits.

The most remote campsite in the national park, the Pool Campground is a park-ranger favourite, set next to a sheltered bay called Smooth Pools. It has room for seven vehicles; facilities are limited to toilets. With room for just two cars, the more exposed Sensation Beach Campground (no facilities) offers easy access to one of the park's loveliest stretches of sand.

The last town before you enter the park is Coffin Bay, renowned for its oysters that can be sampled fresh on a number of tours operated by Experience Coffin Bay (www.experiencecoffinbay.com.au). Oyster season peaks from May to July.

THE PITCH

Choose from a range of secluded campsites in this remote national park at the tip of the Eyre Peninsula, with summertime visits ideal for swimming in its protected bays.

When: Nov–Mar
Amenities: most campgrounds have toilets
Best accessed: by car
Cost: from $21.50
Contact: www.parks.sa.gov.au

10

© RUGLIG | SHUTTERSTOCK

DHILBA GUURANDA-INNES NATIONAL PARK
YORKE PENINSULA

11

Just over an hour's drive from Adelaide, the Yorke Peninsula is one of South Australia's most accessible seaside escapes. But it's worth driving a little further from the city to reach the national park at the peninsula's southern tip. Of the park's five campgrounds, Casuarina is a top choice, with its beautiful bush setting and proximity to the splendid Pondalowie Bay Beach, just a 10-minute walk through the dunes. But the park also offers an opportunity to bed down in the ghost town of Inneston, where six restored historic buildings sleep between two and 10 people. Settled in the early 1900s, the gypsum mining town was abandoned after its plaster factory was forced to close during the Great Depression, after which Inneston was slowly reclaimed by the coastal scrub.

While basic, all of Inneston's lodgings have toilets, showers, filtered rainwater, electricity, stoves, fridges and basic cooking utensils, with linen suppled only at the one-bedroom Post Office Lodge. In the park's northwest, near the Shell Beach Campground and the idyllic Blue Pool rock pool, the rustic Shepherd's Hut invites you to live like Inneston's early pioneers, with no electricity or toilets, though the Shell Beach loos are only a short walk away.

⊖ THE PITCH

Spend an eerie night bedding down in a historic ghost town surrounded by coastal wilderness, or pitch a tent in the Yorke Peninsula's coastal dunes.

When: year-round
Amenities: all campgrounds have toilets
Best accessed: by car
Cost: from $25.50
Contact: www.parks.sa.gov.au

© MYPHOTOBANK.COM.AU | SHUTTERSTOCK

TRIG CAMPGROUND
DEEP CREEK NATIONAL PARK, FLEURIEU PENINSULA

Less than two hours' drive south of Adelaide, Deep Creek National Park has long been a popular destination for a day trip or weekend getaway for city folk. Home to the largest portion of remaining natural vegetation on the Fleurieu Peninsula, it provides a sanctuary for an array of native wildlife such as western grey kangaroos, short-beaked echidnas and 100 species of birds that can be heard and seen while walking in the park. Hardy wintertime visitors are often rewarded with sightings of whales passing the coast during their annual migration between June and October.

Five campgrounds are dotted throughout the park, with four of them accessible by 2WD. While every Adelaidean has their favourite, Trig Campground is a top choice. Situated in the middle of the park, it's got a fabulous edge-of-the-world feel, with sea views, grassy areas popular with kangaroos, and easy access to the best of the park's 15 hiking trails incorporating sections of the Wild South Coast Way on South Australia's famous long-distance Heysen Trail. Each of the 25 camping sites are well spaced between grass trees and eucalypts, though there's not a lot of protection if the wind picks up.

🔴 THE PITCH

Gaze across to Kangaroo Island from your elevated campsite in the rolling hills at the tip of the Fleurieu Peninsula, which you'll likely share with a 'roo or three.

___

When: year-round
Amenities: toilets, picnic shelters, firepits
Best accessed: by car
Cost: from $25.50
Contact:
www.parks.sa.gov.au

12

13

ECOPIA RETREAT
KANGAROO ISLAND

Kangaroo Island's famous wildlife isn't limited to its national parks. Set within a 60-hectare registered wildlife sanctuary in the heart of the island, Ecopia's two one-bedroom Eco Villas offer a unique secluded base for wildlife watching, with a solar passive design allowing stable temperatures to be maintained inside with minimum energy expenditure. Careful thought was also put into the amenities, from the power-saving coffee machine to the eco-friendly bathroom products.

Villas provide guests with an intimate experience, with sweeping views of the surrounds through the wide windows. In a personalised touch, walls are hung with beautiful pieces of Indigenous art. After tucking into your welcome pack of premium local produce (including wine), wander down to the river which flows through the property and spot endangered glossy black cockatoos cackling in the native trees. Stargaze from your deck, and wake up in your king bed to see kangaroos grazing in golden light.

There's also a bookable homestead and a two-bedroom retreat on the property, easily reached by 2WD. With a 35-minute drive to the island's main town of Kingscote, it's best to pick up extra food supplies before you head out here.

🌿 💱 THE PITCH

Enjoy a bit of eco-luxury at this pair of secluded, off-grid villas nestled in a private wildlife sanctuary in the heart of Kangaroo Island.

When: year-round
Amenities: drinking water, electricity, wi-fi, kitchen, barbecue
Best accessed: by car
Cost: $1600, 2-night minimum
Contact: www.ecopiaretreat. com.au

© QUENTIN CHESTER | ECOPIA

MT REMARKABLE NATIONAL PARK
SOUTHERN FLINDERS RANGES

Named for the 'lofty way' the national park's highest peak towers over the landscape, the southern fringe of the Flinders Ranges really is remarkable, with gorgeous gorge hikes, incredible views and abundant wildlife, from the elusive yellow-footed rock wallaby to kangaroos and mobs of emu.

With four entrances to the park, a plan of attack is essential. If you're camping at either of the two main 2WD-accessible campgrounds on the southwestern edge (Baroota and Mambray Creek), take Park Road off the Augusta Highway. The lion's share of the park's trails begin at the day visitor area at the end of this road, close to the campgrounds.

To hike the Mt Remarkable Summit Loop (13.8km/8.5 miles; allow 5hrs) on the park's southeastern fringe before making camp, take the Horrocks Highway to Melrose. The Willowie mountain bike trail network is also off the Horrocks Highway, while Alligator Gorge Road brings you into the top end of the park for excellent hiking including the Alligator Gorge Ring Route Hike (9km/5.6-miles; allow 4hrs) with its spectacular narrow gorge. After a long day on the park's trails, its bush campsites offer a peaceful retreat. Note hike-in campsites are typically closed during the fire danger season (usually November to April).

14

© N MRTGH | SHUTTERSTOCK

🌐 THE PITCH

No less than 15 hiking trails lace this dramatic national park north of Adelaide, with two easily accessible campgrounds just inside the entrance, and an additional 11 hike-in options.

When: Apr–Oct
Amenities: untreated water, toilets, firepits, showers at Mambray Creek
Best accessed: by car
Cost: from $25.50
Contact: www.parks.sa.gov.au

BELLWETHER WINES GLAMPING & CAMPING
COONAWARRA

15

Famous for its vivid red soil – terra rossa – and natural aquifers, the unique geology of South Australia's Limestone Coast was tailor-made for viticulture. While best known for its bold cabernet sauvignons, the region's wineries also bottle everything from merlot to shiraz, chardonnay to cabernet franc.

There are more than 25 cellar doors in the Coonawarra region and you can pitch up right next to one of them at Bellwether Wines. Shaded by 500-year-old red gums, the winery has six unpowered glamping tents and six campsites, two of which are powered. And here's the best bit: all sites are positioned at least 100m (328ft) apart for privacy. The bathroom facilities are shared, as is a semi open-air camp kitchen, with access to a herb garden. Lanterns are supplied for glamping tents and there are power points in the kitchen and bathrooms for charging phones.

The campground is just a short stroll from the cellar door, housed in a beautiful old stone building. There's a whole suite of tasting experiences to choose from, and the UNESCO-listed Naracoorte Caves are just a 20-minute drive up the road. After dark, the communal firepit crackles to life, inviting guests to gather round and sip local wine under the stars.

🛈 THE PITCH

Glamp or camp surrounded by ancient red gums and within stumbling distance of a cellar door in the Coonawarra wine region, known for its exceptional cabernet sauvignons.

When: year-round
Amenities: shared bathrooms, shared kitchen, electricity, drinking water
Best accessed: by car
Cost: camping from $40
Contact: www.bell wetherwines.com.au

GAWLER RANGES NATIONAL PARK
EYRE PENINSULA

The Eyre Peninsula typically evokes images of the sea – sandy beaches, inviting rock pools and gorgeous sunny days by the water. And there's no shortage of any of them. But when the weather cools, it's the perfect time to explore the peninsula's outback scenery in the Gawler Ranges to the north.

A former sheep station, the Gawler Ranges National Park boasts some incredible natural features including the Organ Pipes, among the world's largest volcanic rhyolite formations. The park's sparse vegetation also makes it easier to spot the emus, kangaroos and wombats that roam its rocky, undulating landscapes. Don't forget to look more closely for lizards including the iconic thorny devil.

There are six campgrounds in the national park, with 4WD-only Kolay Campground offering easy access to Kolay Falls, which has its own set of 'organ pipes' that glow a vibrant vermilion at sunset. From the high-clearance 2WD-accessible Waganny Campground, a 90-minute-return trail leads to a spectacular rocky outcrop with 180-degree views, while in a gully offering partial shade, the Yandinga Campground, also accessible by 2WD, positions you closer to the Organ Pipes Walk (one hour). Taking you to a magnificent natural amphitheatre surrounded by 1500-million-year-old rock formations, it's a must-do. On clear nights, all campgrounds morph into million-star hotels.

16

© NICOLE PATIENCE | SHUTTERSTOCK

💬 THE PITCH

After spotting emus and 'roos on a self-guided 4WD safari, camp under stars that feel so close you can almost touch them in this outback park.

When: Mar—Nov
Amenities: toilets, firepits
Best accessed: by car
Cost: from $21.50
Contact: www.parks.sa.gov.au

TASMANIA

The castaway island dream is real in Tasmania, be it camping on beaches or mountains (or even beaches on mountains) or heading to wildlife-filled islands off an island.

Best time: Oct–Apr
Best national parks for camping: Maria Island National Park, Freycinet National Park, Narawntapu National Park, South Bruny National Park
Best camping trails: South Coast Track, Freycinet Peninsula Circuit
National parks pass required: Vehicle pass required for all parks — day passes, holiday passes (two months) and annual passes available.
Useful contacts: www.discovertasmania.com.au; https://parks.tas.gov.au

With 19 national parks covering more than 40 per cent of the island, including the massive Tasmanian Wilderness World Heritage Area that blankets one quarter of the state, Tasmania is all about the outdoors — even Hobart is pinched between a mountain and a river.

Roads often barely penetrate Tassie's protected areas, making camping the finest way to immerse yourself in its wild areas. You can roll out a sleeping mat beside the most famous beach in the land, sleep in virtual sight of Cradle Mountain, or throw down a tent in a different location each night as you progress along one of the multitude of multi-day hikes. Camping here is as much about solitude as a chance to meet the locals — in this case, Tasmania's copious and curious wildlife.

FREE CAMPING

Fees apply in most national park campgrounds that can be reached by road — exceptions include far-southerly Cockle Creek and Bruny Island's Jetty Beach — but there's million-dollar real estate that you can claim gratis at the campgrounds lining the Bay of Fires. Sites found along walking tracks are typically free.

SUPPLIES

Hobart and Launceston are your best bets for any camping gear. Both Mountain Creek Outdoors (Hobart; www.mountaincreekoutdoors.com.au) and Aspire Adventure Equipment (Launceston; https://aspireadventureequipment.com.au) are locally owned outdoors stores.

SAFETY

There's a certain assurance when you see a snake in Tasmania — all three species here are venomous, so be vigilant. Higher roads can get covered by snow in winter (check for closures at www.police.tas.gov.au/community-alerts) and be alert for bushfires during the warmer seasons, especially if camping along walking trails. Fire alerts are posted at https://alert.tas.gov.au.

BEST REGIONS

East Coast

Pick a beach, any beach. Tasmania's east coast would possibly be Australia's top beach destination if only it had the weather to match. Freycinet dangles languidly into the sea, and Maria Island crawls with timid wildlife. And don't get us started on the colours of the Bay of Fires.

Cooking at the Bay of Fires Bush Retreat (left); hiking at Corinna in the Tarkine (below)

Tasmanian Wilderness World Heritage Area

Stretching from Cradle Mountain to the south coast is this mountain landscape cut into pieces by deep, dark rivers such as the famous Franklin and Gordon. This is the place to take your camping wild.

Bruny Island

Forget stodgy camp food. Bruny provides the best of camp kitchens, filled with artisan cheeses, oysters, craft beer, chocolate, berries, whisky and wine. The empty beaches, tall cliffs and remote walking trails are almost just a bonus.

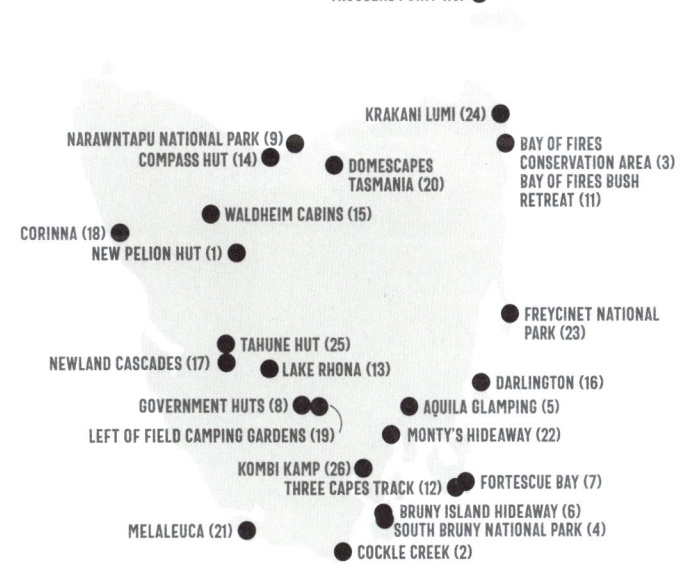

TROUSERS POINT (10)

KRAKANI LUMI (24)

NARAWNTAPU NATIONAL PARK (9)
COMPASS HUT (14)

BAY OF FIRES
CONSERVATION AREA (3) &
BAY OF FIRES BUSH
RETREAT (11)

DOMESCAPES
TASMANIA (20)

WALDHEIM CABINS (15)

CORINNA (18)
NEW PELION HUT (1)

FREYCINET NATIONAL
PARK (23)

TAHUNE HUT (25)
NEWLAND CASCADES (17)
LAKE RHONA (13)

DARLINGTON (16)

GOVERNMENT HUTS (8)
LEFT OF FIELD CAMPING GARDENS (19)

AQUILA GLAMPING (5)

MONTY'S HIDEAWAY (22)

KOMBI KAMP (26)
THREE CAPES TRACK (12)

FORTESCUE BAY (7)

BRUNY ISLAND HIDEAWAY (6)
SOUTH BRUNY NATIONAL PARK (4)

MELALEUCA (21)

COCKLE CREEK (2)

NEW PELION HUT
CRADLE MOUNTAIN-LAKE ST CLAIR NATIONAL PARK

Midway along Australia's most famous mountain trail, the Overland Track, New Pelion Hut is akin to a crossroads inn. It's here, more than at any other hut along the route, that Overland Track hikers often pause for a day of rest, while others traipse into the hut from along the lesser-known Arm River Track.

This beautiful bottleneck of hikers and mountains is larger than most other huts along the Overland Track, with a spacious shared area. It also overflows with distractions. The front deck peers across the buttongrass-covered Pelion Plains to the dramatic figure of Mt Oakleigh, with its dolerite columns rising like horns. From the hut, a trail ascends to the top of this mountain, while New Pelion is also a springboard to summit climbs of Mt Ossa (Tasmania's highest mountain), Mt Pelion East and, further afield, the likes of Mt Pelion West and mythologically named Mt Thetis and Mt Achilles. Closer at hand, just a short walk from its replacement, is Old Pelion Hut, built in 1917 to house copper miners. It's now a day-use shelter with a palpable sense of history.

Behind New Pelion Hut is its camping area, with a tent platform large enough to hold three tents and overflow camping on the surrounding grass.

🪨 THE PITCH

The perfect pause along the Overland Track, surrounded by attainable summits and historical sites, with room to sleep 36 hikers.

When: Oct–May
Amenities: pit toilets, tank water
Best accessed: by hiking only
Cost: included in the $285 Overland Track Pass
Contact: https://parks.tas.gov.au

01

© RYAN HOI | SHUTTERSTOCK

COCKLE CREEK
SOUTHWEST NATIONAL PARK

02

⚡ THE PITCH

World's-end camping in Southwest National Park's most accessible campground, around two hours' drive south of Hobart.

When: Oct–Apr
Amenities: pit toilets, tank water, payphone
Best accessed: by car
Cost: free
Contact: https://parks.tas.gov.au

Drive south. As far south as you can go, to where the road ends. Only then will you arrive at Cockle Creek, at the edge of the Tasmanian Wilderness World Heritage Area and seemingly also the edge of the world. To get any further south than this you must walk, and beyond that there's only the ocean and Antarctica.

The main access point for Tasmania's largest national park, Cockle Creek is a campground of two halves. On one side, just before you enter the national park, the numbered campsites in Recherche Bay Nature Recreation Area allow dogs and fires (in firepits) and tend to draw a louder crowd.

Step over the national park border and things settle. Camping here is more along traditional national-park lines: no pets, fuel stoves only, weary walkers with early bedtimes. There are around a dozen sites on the bush-fringed lawns, with additional spots for self-contained RVs just beyond.

The unstaffed park visitor centre marks the end (or start) of the week-long South Coast Track. The first part of this remote trail makes for a popular day walk to South Cape Bay, where you can see Australia's southernmost point, South East Cape, over the top of the pounding Southern Ocean surf.

03

BAY OF FIRES CONSERVATION AREA
BAY OF FIRES

Imagine million-dollar real estate, free of charge, a short walk from some of Australia's most beautiful beaches. The Bay of Fires was once named the world's hottest travel destination by Lonely Planet, and yet a string of simple campgrounds remains its only beachfront accommodation.

From Binalong Bay, a surprisingly quiet road leads to the tiny settlement of The Gardens, home to seven first-come, first-served campgrounds, almost all of which have access to magical beaches.

The first campsite, Grants Lagoon, sits back from the ocean, but Jeanneret Beach, Swimcart Beach, Cosy Corner South, Cosy Corner North, Seatons Cove and Sloop Reef step straight out onto beaches of flour-white sand, brilliantly blue seas and trademark blazes of orange lichen across granite boulders and headlands. It's the most colourful stretch of coast in the state.

Naming the best campsite is like choosing a favourite child, though Swimcart Beach is the largest. It's also the finish post for one of Tasmania's best mountain bike trails, the Bay of Fires Trail.

🧭 THE PITCH

A string of seven basic campgrounds along one of Tasmania's most beautiful slices of coast. Pick a beach, any beach.

When: year-round
Amenities: pit toilets (no toilets at Seatons Cove and Sloop Reef); bring your own drinking water
Best accessed: by car
Cost: free
Contact: https://parks.tas.gov.au

© TRAVELNERD | SHUTTERSTOCK

SOUTH BRUNY NATIONAL PARK
BRUNY ISLAND

Bruny Island is a delicious destination, in the most literal of senses. Spread like a food platter in the sea, it dishes up artisan cheese, oysters, craft beer, honey, chocolate, whisky, freshly baked sourdough from retro roadside fridges, and wine from Australia's southernmost vineyard. Drive on past temptation to Bruny's southern shores and it's also a camping destination par excellence.

Tucked into a pocket of Labillardiere Peninsula, the Jetty Beach campground backs onto a beautiful and protected north-facing beach, away from the Southern Ocean swells, that's great for swimming. One bay over, in spectacular and wild Cloudy Bay, ignore the very basic Pines Campground just before the beach and set your sights on Cloudy Corner, at the far end of the beach from the road end. You can 4WD along the sand to the campground, or throw on a backpack and walk the 3km (1.8 miles) along the wave-lashed beach into camp.

Both campgrounds sit at the start of rewarding walking trails. From Jetty Beach, you can make an 18km (11-mile) loop of the Labillardiere Peninsula, while it's a 3km (1.8-mile) hike from Cloudy Corner out to remote East Cloudy Head and brilliant views over Cloudy Bay and Cape Bruny.

THE PITCH

Find your own version of gourmet camping in a national park with easy access to some of Tasmania's finest food and wine — it's the best of both worlds.

When: Oct–Apr
Amenities: untreated water, picnic tables, pit toilets, fire rings (at Jetty Beach)
Best accessed: by car
Cost: free
Contact: https://parks.tas.gov.au

04

AQUILA GLAMPING
RICHMOND

From the decks of Aquila's glamping tent and pods, your eyes will be naturally drawn to the sky. This retreat, perched on a hillside above famously pretty Richmond and the vineyards of the Coal River Valley, is named for the endangered Tasmanian wedge-tailed eagle – *Aquila audax fleayi* – seen so reliably in the sky. But this massive bird isn't the only extraordinary sight at Aquila.

From the tent's deck, with its outdoor tub and Adirondack chairs, the view stretches across farmland to a sandstone cliff. This cliff was quarried in the early 19th century to build Richmond Bridge, Australia's oldest stone arch bridge, just 2km (1.2 miles) from Aquila.

Sheep trim the lawns around the tent and shipping-container pods (with double glazing and six-star energy ratings) set on a 20-hectare parcel of farmland. Birds are a recurring motif, from the trio of aviaries to the guinea fowl that wander the small plot of vines and the selection of metal sculptures dotted around the pods.

The tent – one of Tasmania's largest glamping tents – is the star, with its deck-top bath, king bed, wood stove, Huon pine bedhead and built-in kitchen and bathroom, though the pods sit higher up the slopes, commanding views all the way to kunanyi/Mt Wellington.

THE PITCH

Off-grid but accessible farmland retreat 30 minutes from Hobart, with a glamping tent and three pods near Richmond and the Coal River Valley's cellar doors.

When: year-round
Amenities: outdoor bathtub (in the tent and one pod), private bathroom with hot shower, full kitchens in pods (microwave only in tent), wood stoves
Best accessed: by car
Cost: from $300
Contact: www. aquilaglamping.com.au

05

© AQUILA GLAMPING

BRUNY ISLAND HIDEAWAY
ALONNAH

06

Connected to the Tasmanian mainland only by a ferry, Bruny Island is already delightfully removed from everyday life, but this ingeniously designed tiny house takes it a step further.

The off-grid Hideaway sits encircled by bush on a sprawling property under a conservation covenant, with an angular design making clever use of its 28 sq m (300 sq ft) space. All furniture (except the low table) is built into the space, while the elevated ceiling gives the loft bedroom an unexpected roominess. The inviting feature of the long Aurora Window hints at the possibility of seeing the Aurora Australis from bed, and looks down onto marsupial-nibbled lawns – you'll likely see dozens of wallabies or perhaps even rare Tasmanian birds such as the swift parrot and forty-spotted pardalote.

With sunlight pouring through the skylight, the absence of curtains and two walls of sliding glass doors, a stay here can be near enough to living in the open air. East- and west-facing decks allow guests to revel in both sunrises and sunsets, with the Hideaway's pièce de résistance suitably hidden within the afternoon deck – pop open a hatch and a clever sunken bathtub is revealed.

🛏 THE PITCH

A well-named island retreat with a neat array of tiny-house tricks, from a secret bathtub to bedside Southern Lights viewing.

When: year-round
Amenities: hot shower, wood heater, running water, kitchen
Best accessed: by car
Cost: from $385 for 2 nights
Contact: www.airbnb.com.au

07

FORTESCUE BAY
TASMAN NATIONAL PARK

Few campgrounds are as beloved among Tasmanians as Fortescue Bay. Set behind one of the few beaches inside Tasman National Park, it's a holiday favourite, just 90 minutes' drive from Hobart and at the heart of the Tasman Peninsula's outdoor offerings.

The long beach has excellent swimming and snorkelling, and Fortescue's campground is the hub for several of the park's best bushwalks. The multi-day Three Capes Track finishes beside the camp, the final 5km (3 miles) of which is a popular day walk to the heady tip of Cape Hauy for the downstairs view of the cape, its fur seals and towering cliffs.

The Port Arthur Historic Site is also just a half-hour drive away.

The campground is split into two parts – Banksia (26 sites) and Mill Creek (23 sites). The former is the scenic pick of the pair, with its spacious and sandy tent sites beneath lofty stringybark trees a few metres from the beach. Mill Creek, behind the boat ramp, is geared towards campervans and campers with boats.

Fortescue's popularity means that it's one of the few Tasmanian national park campgrounds that requires bookings, at least from November to April (it's first-come, first-served from May to October).

THE PITCH

A perennial beachside camping favourite at the end of a long dirt road, at the heart of the spectacular Tasman Peninsula.

When: Oct–Apr
Amenities: hot showers, pit toilets, gas BBQs, fireplaces, store (selling basic supplies)
Best accessed: by car
Cost: $13
Contact: https://parks.tas.gov.au

© RYAN HOI | SHUTTERSTOCK

GOVERNMENT HUTS
MT FIELD NATIONAL PARK

There aren't many ways to spend a night high on the peaks of Tasmania's oldest national park. Along the alpine walking trails of Mt Field National Park, there are no campgrounds and no hiker huts. But there are the Government Huts.

Sitting by the roadside in a snow-gum stand high above sea level, these five rustic timber cabins were originally built to provide accommodation for workers on this road that climbs through the mountains to Lake Dobson, and later moved to their present location.

The huts are frill-free – think camping with wooden walls – containing bunks with mattresses, tables, benches and wood heaters, but no electricity. Toilets are shared, and you need to bring your own sleeping gear. For all that, the huts are also a rare all-access ticket into Mt Field's alpine regions.

A ten-minute walk beyond the Government Huts, Lake Dobson is the trailhead for many of Mt Field's best walks. Round the lake to enter the otherworldly Pandani Grove, home to the endemic tree that looks like a Dr Seuss creation, or set off up the slopes to reach the Tarn Shelf. This string of small mountain lakes is famous for the autumn turning of the fagus, when Australia's only native winter deciduous tree turns golden.

08

⊖ THE PITCH

Head for rare air – 1000m (3280ft) above sea level – without having to hike up to your hut, at these basic workers' cottages turned popular mountain bases.

When: Oct-Apr
Amenities: mattresses, running water, tables, wood heater
Best accessed: by car
Cost: hut $45
Contact: https://parks.tas.gov.au

NARAWNTAPU NATIONAL PARK
NORTH COAST

09

It's a stretch to liken the plains of Narawntapu National Park to the African savannah, but it only takes a stroll across the lawns at dusk to appreciate why this park has been dubbed the Serengeti of Tasmania. The Springlawn area at Narawntapu's heart is one of the most densely packed macropod hangouts in the state, filling each day with Forester kangaroos, Bennett's wallabies and pademelons.

Narawntapu's principal campground is right in the thick of the marsupial action – you'll likely have more wallabies than people as neighbours in this cul-de-sac of grassy powered sites – and it's a short walk to Springlawn Lagoon, where you'll find a bird hide for spying on the many ducks, swans and grebes.

Most of the park's walking tracks radiate out from Springlawn, making it the perfect base for days wandering long Bakers Beach, climbing Archers Knob, or crossing the entire park on the full-day, 21km (13-mile) coastal traverse to Greens Beach.

In addition to Springlawn, there are two other excellent campgrounds backed up against the Rubicon Estuary. Koybaa is a short loop with 12 small tent sites, while the larger Bakers Point (32 unpowered sites) makes for a very short walk out to the water and beach. Naturally, they all come with wallaby company.

🐾 THE PITCH

Wildlife viewing doesn't come easier than at this trio of coastal national park campgrounds an hour's drive northwest of Launceston.

When: year-round
Amenities: hot showers, untreated water, flush toilets (Springlawn), pit toilets (Koybaa, Bakers Point)
Best accessed: by car
Cost: $16/$13 powered/ unpowered
Contact: https://parks.tas. gov.au

10

© ALEX CIMBAL | SHUTTERSTOCK

TROUSERS POINT
STRZELECKI NATIONAL PARK, FLINDERS ISLAND

If you've only seen one image of Flinders Island, it's likely to be the curiously named Trousers Point. This island is like Tasmania in microcosm – imposing mountains, blazing lichens on granite headlands, plentiful wildlife and an abundance of hiking trails – yet Trousers Point stands out as the beauty among beauties. Its crescent-shaped beach is framed by boulders seemingly dipped in orange lichen, with the bare slopes of the 782m (2566ft) Strzelecki Peaks rising behind.

Just before the sands is the Trousers Point campground. This basic camp has a location that every other accommodation provider on the island can only envy. It's just a few steps to the island's most famous beach, where, on a calm day – not a given on this island that straddles the infamous Roaring 40s winds – you can plunge in to snorkel the sheltered shores. The popular Trousers Point walk also begins (and ends) at the campground. This walk, running from the small point to the equally beautiful Fotheringate Beach, is one of the three Tasmania's Great Short Walks on Flinders Island, along with the climb to the summit of the Strzelecki Peaks; the start of this track is just 4km (2.5 miles) away.

🧭 THE PITCH
An island off an island, with a simple but spectacular national park campground right beside one of Tasmania's most beautiful beaches.

When: Nov-Apr
Amenities: gas BBQ, tank water, pit toilets, picnic tables, fireplace
Best accessed: by car
Cost: free
Contact: https://parks.tas.gov.au

BAY OF FIRES BUSH RETREAT
BINALONG BAY

There's an unmistakable chef's touch to Bay of Fires Bush Retreat – standing proud among its 10 bell tents is a large, semi-open full kitchen. Pre-prepared meals from a local chef – the likes of slow-cooked lamb ragout or harissa chicken – fill a fridge, and there's an honesty bar brimming with Tasmanian cheeses, beers and wines. Missing a dinner ingredient? Forage through the sizeable seasonal vegetable garden for herbs, kale, lemongrass, potatoes or summer berries.

Owned by prominent Tasmanian chef, Tom Dicker, and partner Anna Hoffmann, the retreat is just 2km (1.2 miles) from Binalong Bay, the beach town at the southern end of the Bay of Fires. Each tent contains a king bed, and the kitchen promotes a sense of communality, reinforced around the shared firepit.

Amenities are also shared and semi-open – peer into the treetops as you shower – and it's just a few minutes' drive to some of Tasmania's best beaches along the dazzling shores of the Bay of Fires. Turn your sights inland and there's the rainforest-draped Blue Tier mountains, waterfalls such as 90m (295ft) St Columba Falls, one of Tasmania's highest, and cheesy goodness at Pyengana Dairy. Or you could sip a beer with a pig at Pyengana's quirky Pub in the Paddock, 30 minutes' drive from the retreat.

11

THE PITCH

Find glamping with a fine-food focus at this bush-clad bell tent retreat, set on Tassie's most beautiful string of beaches.

When: mid-Aug–mid-Jul
Amenities: hot showers, full kitchen, wi-fi, running water, flush toilets, firepit
Best accessed: by car
Cost: from $220
Contact: https://bayoffires bushretreat.com.au

© ADAM GIBSON

THREE CAPES TRACK
TASMAN NATIONAL PARK

In 2015, a new benchmark was set for public huts with the launch of the Three Capes Track along the tops of the country's highest sea cliffs. Three huts were strung along its 48km (30-mile) length – but these were no ordinary huts. Gone were the bare-bones wooden platforms and stainless-steel tables, upgraded to the likes of mattresses, USB ports, yoga mats, canvas deck chairs, reference libraries and pots, pans and cutlery.

The well-equipped Surveyors, Munro and Retakunna huts improve the comfort of both nights and days on the track, cutting down on the gear hikers need to carry, leaving room in backpacks for other essentials such as, um, wine and higher-quality food. Surveyors and Munro huts come with million-dollar views: Surveyors, the first night's stop, looks across the water to dramatic Cape Raoul, while the deck at Munro commands cinematic views over Munro Bight to rocky and rugged Cape Hauy.

Each hut has a mix of four- and eight-bunk rooms with memory-foam mattresses. The separate hub buildings feature passive solar design, pellet heaters burning a timber waste by-product, and windows sloped to minimise bird strike. Each hut is staffed by a host ranger, like a free-range concierge.

THE PITCH

One of Australia's most spectacular coastal hikes, capped with a trio of the country's plushest public huts.

When: year-round
Amenities: gas stoves, mattress beds, untreated water, USB ports, compost toilets, hot showers (at Munro hut)
Best accessed: by hiking only
Cost: $595 to hike the track, including three nights in the huts
Contact: www.threecapestrack.com.au

12

© JIRI VIEHMANN | SHUTTERSTOCK

Paddling adventures

Home to a river trip described by *Outside* magazine as the world's best white-water rafting journey, Tasmania can truly float your boat when it comes to paddling adventures.

Wrapped in ocean and veined with rivers, Tasmania is a water world of aquatic opportunities. Be it a gentle kayak trip into Hobart's docks for a floating feed of fish and chips, or a multiday epic down the Franklin River, ocean and river journeys with paddle in hand have been a part of the Tasmanian outdoor psyche ever since John Dean, John Hawkins, Trevor Newland and Henry Crocker first canoed the Franklin – at the third attempt, six years after their first try – in 1958.

These are waterways so dear to hearts that they fostered Australia's Green movement – Bob Brown, the most famous leader of the Australian Greens political party, found his passion for the environment when he became the first person to raft the Franklin in 1976. Today, rafts continue to bump through gorges and rapids on rivers such as the King and Huon, while kayak trips explore coastlines and offshore islands.

THE FRANKLIN BLOCKADE

In the early 1980s, plans were afoot to dam the remote Franklin River, leading to a protest and blockade on a scale never before seen in Australia. More than 1200 protestors were arrested across three months, and Bob Hawke was elected as Prime Minister on a promise to save the river, a move that led to the creation of the Tasmanian Wilderness World Heritage Area and secured the Franklin as one of Australia's most famous and cherished waterways.

SELF-PROPELLED ON THE MEANDER

Not all paddling trips involve paddles. The Meander River in Tasmania's north is one of only two places in Australia with guided river-sledding trips. On these outings, participants lie down on lilo-like sleds, steering and paddling with their hands through grade I and II rapids.

Rafting Thunderush rapid on the Franklin River (top), tranquil Pieman River in the Tarkine (above); Bramble Cove in southwest Tasmania and loaded kayaks (inset)

FIVE TO TRY

01

Raft the Franklin River
Australia's ultimate rafting adventure, going deep into pristine wilderness. Trips typically take eight days.

02

Kayak at Corinna
Hire a kayak and set out on the mirror-like Pieman River to Lovers Falls and an inland shipwreck.

03

Kayak through Bathurst Harbour
Weeklong trips feel utterly remote, not a short jaunt from Hobart.

04

Packraft the Mersey River
Inflate a small packraft for a three-day guided journey through beneath the magnificent Alum Cliffs.

05

Kayak the Tasman Peninsula
Set off to the cliffs of Cape Hauy, or on a seasonal trip among migrating whales.

LAKE RHONA
FRANKLIN-GORDON WILD RIVERS NATIONAL PARK

Usually when camping you're forced into a choice between beach and mountains, but Lake Rhona has you covered for both. Benched into the remote and roadless Denison Range, the lake was once a closely guarded local secret, but recent years have seen its popularity grow to the point that a walker registration system has been implemented – be sure to book your lake stay in advance.

It's hard work to get here, with a full day of walking that includes crossing the Gordon River on the trunk of a fallen tree (only possible when water levels aren't high), long muddy stretches through buttongrass moors, and a final 400m (1312ft) ascent onto the range. First sight of the tannin-darkened lake is enough to reward the slog, with Lake Rhona partly rimmed by a long, pink-tinged quartzite beach.

Camping here is a primitive and ad hoc affair, with tents pitched on the sands or in sites tucked more protectively into the backing bush. It's a stunning location, with the tea-coloured water often reflecting the sharp peaks of the Denison Range that rise like thorns above.

The next morning it's possible to follow a trail up onto the range to the summit of craggy Reeds Peak. Or you might just prefer that rarest of mountain experiences – some beach time.

13

⊘ THE PITCH

Hike to a mountain marvel to discover the curiosity of beachside camping high on the slopes of a rugged range.

When: Oct–Apr
Amenities: pit toilet
Best accessed: by hiking
Cost: free
Contact: https://parks.tas.gov.au

COMPASS HUT
FORTHSIDE

14

Off-grid, organic and overlooking Bass Strait, Compass Hut might feel remote and rural, but Tasmania's Devonport and the *Spirit of Tasmania* ferry are just a 15-minute drive away.

Rolled onto the slopes of a large family farm are three tiny houses cleverly designed by their owners, sisters who grew up on the property. Standing up to 500m (1640ft) apart, with the farmhouse tucked away over a hill, each house is distinct and discrete. The oldest, Barnhaus, is also the smallest, and is modelled on a Scandinavian barn, with a double bed and an outdoor deck for dining. The larger Arc Pavilion, with its outdoor bath, is wrapped in decks and glass walls to absorb the sun, and has bifold doors to bring the outside in. A swinging interior wall can also divide the single room. Newest in the collection is Colonial Blue, a timber-lined tiny house with a separate ensuite containing a double shower and bathtub. It doubles as an exhibition space, and showcases artworks from the owners and other artists.

Sustainability is a big factor in Compass Hut's appeal, and each tiny house is set in private gardens, including an edible one. There are also rainwater collection, compost toilets and solar panels feeding into Tesla Powerwall batteries.

⊘ 🝙 THE PITCH

Organic existence on an organic farm, with three tiny houses on wheels strung across the slopes of a working property.

When: year-round
Amenities: hot showers, outdoor bath (in Arc Pavilion), indoor bath (in Colonial Blue), running water, kitchenettes, wi-fi
Best accessed: by car
Cost: $200-$320
Contact: www.compasshut.com.au

© KYLIE BELL

WALDHEIM CABINS
CRADLE MOUNTAIN–LAKE ST CLAIR NATIONAL PARK

When Gustav and Kate Weindorfer – names synonymous with Cradle Mountain-Lake St Clair National Park – first climbed Cradle Mountain in 1910, the former famously declared that the place 'must be a national park'. It was a goal achieved, largely through the Weindorfers' lobbying efforts, in 1922, 10 years after the Austrian-born Gustav built a wooden chalet on the edge of Cradle Valley that the couple named Waldheim. It was the first accommodation in the Cradle Mountain area.

Today, on the edge of the myrtle forest behind that chalet, eight cabins continue the tradition. The parks-run Waldheim Cabins are about location, not luxury. Each sleeps between four and eight people and features bunk beds, a basic kitchen and a shared toilet and shower block. You will need to bring your own bedding. However, it's the only accommodation (other than hiking huts) within the national park rather than on its fringe.

The cabins are 5km (3 miles) inside the park, in easy reach of Dove Lake and its celebrated view of Cradle Mountain. You can honour the history behind your stay with a visit to Waldheim Chalet, a replica of the Weindorfers' chalet in its original position beside the entrance to the cabins.

🛏 THE PITCH

A rare opportunity to stay inside Cradle Mountain-Lake St Clair National Park, far from the crowds and near the park's namesake mountain.

When: year-round
Amenities: stove, hot showers, flush toilets, electric heating, fridge; BYO bedding
Best accessed: by park shuttle, car
Cost: $140–$280
Contact: https://parks.tas.gov.au

15

16

© DAVID LADE | SHUTTERSTOCK

DARLINGTON
MARIA ISLAND NATIONAL PARK

Imagine a place where you can camp inside a World Heritage Site, beside convict ruins and among a crowd of wombats, with a likely cameo from a Tasmanian devil or two. Welcome to Darlington.

Behind Darlington's beach, the large, tent-only campground is just steps from the island's convict penitentiary. Wander through its buildings (the oldest – the Commissariat Store – dates to 1825) and venture out on hikes to the brilliantly patterned Painted Cliffs and the well-named Fossil Cliffs, or climb the heady Bishop and Clerk Track and Mt Maria, the island's tallest mountain. Or,

bring a mountain bike and explore Maria's traffic-free gravel roads. Expect to encounter wombats, wallabies and Cape Barren geese wherever you go, or spot one of the Tassie devils successfully introduced to the island.

There are no shops or services on Maria, which is entirely national park, so bring everything you need with you on the ferry

from Triabunna, and then haul it in carts (or by hand) to the campsite (about 500m/0.3 miles from the ferry dock).

If you don't fancy pitching a tent, there's the option of sleeping inside the convict penitentiary, where 10 former cells offer basic visitor accommodation – bunk beds and wood heater, but no electricity.

🔵 THE PITCH

Pitch a tent on a traffic-free prison island turned national park, so abundant in wildlife that it's often been proclaimed as Australia's Noah's Ark.

When: Oct–Apr
Amenities: gas BBQs, fireplace, flush toilets, untreated water
Best accessed: by ferry
Cost: $13
Contact: https://parks.tas.gov.au

NEWLAND CASCADES
FRANKLIN-GORDON WILD RIVERS NATIONAL PARK

17

It's no easy task to reach Newland Cascades. To get to this remote cliffside camp, you must raft the Franklin River, a demanding, wild journey of up to 10 days, described by US magazine *Outside* as the world's best white-water journey. You'll squeeze through the Great Ravine, a gorge so fierce that most of its rapids need to be portaged, bump through Propsting Gorge and emerge into the calm of Rock Island Bend. Only then, about six days down the river, do you splash through Newland Cascades, the Franklin's longest rapid.

As the rapid spits you out, cliffs rising from the right bank provide an unlikely setting for the river's best and most spacious campsite. Shallow caves in the cliffs, and rock platforms outside of them, create spaces in which to roll out your sleeping bag and mat, protected from the all-too-regular rain by the overhangs. Other rocks around the foot of the cliffs create excellent kitchen spaces.

If the river level is low, there's the possibility of scrambling the short distance back to Rock Island Bend, a single photo of which helped turn national sentiment against a plan to dam the Franklin River in the early 1980s.

🧭 THE PITCH

Take one of the world's great rafting trips to a cliffside camp above a lengthy river rapid, sleeping in caves or out on rock ledges.

When: Oct–April
Amenities: none
Best accessed: by white-water raft
Cost: free
Contact: https://parks.tas.gov.au

© AUSCAPE | GETTY IMAGES

18

CORINNA
TAKAYNA/TARKINE

Where once there was gold, there's now only green at Corinna. Home to 2500 people during a late-19th-century gold rush, it today slumbers as a low-key tourist centre at the southern end of takayna/Tarkine, Australia's largest cool-temperate rainforest.

The gold-rush wooden cottages have been transformed into accommodation, but it's campers who get the best of the rainforest real estate. Seven tent sites and three van sites are strung along the bank of the Pieman River, the dark waterway that marks the rainforest's southern edge.

Tasmania's northernmost Huon pines grow from the riverbank, including an ancient beauty just a few hundred metres downstream, and walking trails head through the forest and up to the summit of low Mt Donaldson for grand views. On the water, the *Arcadia II* – the only Huon pine river cruiser still operating in the world – sails downstream from Corinna to wild Pieman Heads, or you can hire a kayak and paddle downstream to quixotic Lovers Falls, surrounded by enormous ferns, or over the remains of Australia's furthest-inland shipwreck.

Corinna is a *palawa* (Tasmanian Aboriginal) name for a young Tasmanian tiger, and there are still people who swear that if the thylacine does live on, it's somewhere around here.

🌀 THE PITCH

Riverside camping at the edge of one of the world's largest tracts of temperate rainforest, surrounded by walks and water activities.

When: Oct–Apr
Amenities: pub/restaurant, coin-operated showers, flush toilets, running water, BBQ
Best accessed: by car
Cost: $40
Contact: www.corinna.com.au

© CHRIS CRERAR

LEFT OF FIELD CAMPING GARDENS
MT FIELD NATIONAL PARK

Only separated from Mt Field National Park by a river, camping at Left of Field is akin to being in a national park, but with a multitude of perks. Stretched along the Tyenna River, its 83 sites (33 powered) are dotted through a garden its horticulturalist owner spent 12 years crafting. It shows, with more than 3500 plant varieties – endemics, exotics and natives – and 65,000 planted bulbs.

Several features distinguish pet-friendly Left of Field from a typical caravan park. There are two outdoor baths, both looking onto the gardens and river. The park has its own quirky nine-hole golf course and one of Tasmania's largest licenced bars, serving more than 200 whiskies and 60 gins from across the world – and you don't have to drive home afterwards. Friday nights bring live music, and each February it hosts the four-day Playing the Field festival, one of the state's longest-running music events.

The Tyenna River is one of Tasmania's top trout waterways and you can almost cast a line from your campsite. There are also regular platypus sightings. Inside the national park, there are walks to waterfalls, tarns and summits, with a night hike to Russell Falls bringing the chance to see glow worms – a more low-key kind of nightlife.

19

⬤ THE PITCH

National-park-edge camping without national park restrictions – pet-friendly, with a lively bar, live music and trout fishing in a garden setting.

When: Oct-Apr
Amenities: bar, hot showers, flush toilets, fireplaces in most campsites, communal firepit, gas BBQ
Best accessed: by car
Cost: $15
Contact: www.leftoffield.net.au

DOMESCAPES TASMANIA
TAMAR VALLEY

Domescapes Tasmania's geodesic domes are all about the 'viner' things in life. This trio of glamping domes sit directly beside the vines of Swinging Gate Vineyard, and just a few minutes' stroll from the cellar door – no designated driver needed on this stay.

The large domes, spaced around 70m (230ft) apart, each feature a king bed with ensuite and shower. Around 25% of the dome's surface is a transparent 'sky window', allowing for stargazing from bed and the embrace of sun-lit Tamar Valley days. Warmth is guaranteed, with insulated walls and underfloor heating, and each dome has its own firepit. Two of the domes also have a clawfoot bath on the deck in which to perhaps enjoy a glass of bubbles in the bubbles.

Each booking receives a bottle of Tamar Valley wine and a local cheese platter, but Swinging Gate's quirky, farmhouse-rustic cellar door always beckons – enjoy a tasting with Nellie, the winery's attention-stealing Staffy, who once featured in a *Wine Dogs of Australia* book.

The bulk of the Tamar Valley's vineyards, as well as Launceston and Narawntapu National Park, are within a 45-minute drive. And watch this space – a fourth dome large enough for families is coming in 2024.

🍽 THE PITCH

Glamping domes inside a Tamar Valley vineyard with grapes growing almost to the front step, and a short stroll to the cellar door.

When: Oct-Apr
Amenities: hot shower, flush toilet, running water, outdoor bath, minifridge, toaster
Best accessed: by car
Cost: $295
Contact: www. domescapes.com.au

20

© DOMESCAPES TASMANIA

21

MELALEUCA
SOUTHWEST NATIONAL PARK

To call Melaleuca a settlement is to inflate the truth, despite what maps might suggest. Once a one-man tin-mining 'town', it's now the heart of Southwest National Park. The only ways into the far-flung site are by air, boat or days of walking through remote wilderness – little more than 100km (62 miles) from Hobart, this remains one of the most isolated places in Australia.

Two green Nissen huts and a campground at the edge of Melaleuca Lagoon are its overnight options. The huts were built for hikers by the legendary Deny King, who lived and mined at Melaleuca for 55 years.

Between them, the spartan huts sleep 20 people in bunks, with one featuring a stone fireplace and a dining table made from Tasmania's celebrated Huon pine. The small campground has makeshift sites strewn among the trees.

Melaleuca might easily have become a ghost settlement, if it wasn't the meeting point for two long-distance bushwalking tracks: the little-trodden Port Davey Track and the popular South Coast Track. Visitors can check out the Deny King Heritage Museum near the airstrip, wander the short Needwonnee Walk, and head to the bird hide for the chance to sight one of the world's rarest birds, the critically endangered orange-bellied parrot.

🐾 THE PITCH

Go remote – no, further than that – to find a pair of Nissen huts and a lagoon-side campground in the middle of a vast wilderness.

When: Nov-Apr
Amenities: pit toilets, untreated water
Best accessed: by plane, hiking
Cost: free
Contact: https://parks.tas.gov.au

MONTY'S HIDEAWAY
HOBART

What if we told you that you can have a night under the stars and still be just a 15-minute walk from Hobart's CBD? That you can sleep in the open air and still grab your morning coffee from a city lane?

Hidden at the bottom of the garden of a heritage Georgian home is this converted shipping container with a distinctive party trick – Monty's Hideaway's roof retracts, pulling back like stage curtains so that the only thing between you and the night sky is atmosphere. The timber-panelled container is as cosy as a tiny home, with the bed positioned to maximise the starry view, underfloor heating, smart TV with streaming (if that natural screen overhead isn't enough), outdoor fire and Bluetooth speaker.

The property on which the container is located was built in 1837 within Hobart's original dress circle, where the city's wealthiest residents established their homes. Immediately opposite is Fitzroy Gardens, a beautiful parkland that puts on arguably the city's most golden autumnal display each year. The cafes and restaurants of chichi Sandy Bay are only a few minutes' walk away, and it's just one block to join the main driving route to the summit of kunanyi/ Mt Wellington, the peak that dominates the city.

THE PITCH

Roll back the roof and sleep under the stars in the heart of Hobart – it's part-camp but full comfort.

When: year-round
Amenities: hot shower, underfloor heating, bar fridge, hair dryer, kettle, toaster
Best accessed: by car
Cost: $150–$230 for 2 nights
Contact: www. bishopsquarters.com.au/ hideaway

22

© SHIII6 | SHUTTERSTOCK

FREYCINET NATIONAL PARK
EAST COAST

Tasmania's most-visited national park owes most of its fame to a single superstar beach: Wineglass Bay. To get to this shapely curve of sand, you must walk, following the Freycinet Peninsula Circuit through a low pass in the Hazards mountains and down onto the bay's gleaming white sands. If you're prepared to carry in camping gear, you can pitch a tent in a small walkers' campsite at the beach's far end, mingling with wallabies and discovering the bay on an intimate scale. You'll need to bring your own drinking water.

The park's more popular camping option is Richardsons Beach, an elongated campground lining a protected beach. With 18 powered sites for caravans and campervans to one side, and 27 tent-only sites atop dunes on the other, it provides absolute waterfront camping. Such is Richardsons' popularity, it's the only Tasmanian campground that operates a ballot system to secure sites across the summer school holidays and Easter.

Wineglass Bay is in easy reach, but be sure to look beyond this one dazzling strand – Freycinet is not a one-trick park. Stroll out to Cape Tourville, where whales are often sighted, take in a sunrise at Sleepy Bay, and scale Mt Amos for one of Tasmania's best coastal views.

THE PITCH

Set up camp within easy reach of Wineglass Bay at this ever-popular national park, or hike to a basic campsite behind the famous beach itself.

When: Sep-May
Amenities: showers, flush toilets and drinking water (Richardsons Beach); pit toilets (Wineglass Bay)
Best accessed: by car (Richardson Beach); hiking only (Wineglass Bay)
Cost: $13/$16 unpowered/ powered (Richardsons Beach); free (Wineglass Bay)
Contact: https://parks.tas. gov.au

23

© NIGEL KILLEEN | GETTY IMAGES

KRAKANI LUMI
MT WILLIAM NATIONAL PARK

When English explorer Tobias Furneaux sailed along Tasmania's northeast coast in 1773, he observed so many Tasmanian Aboriginal fires burning along its shores that he named the area the Bay of Fires. On evenings at krakani lumi today, it's like history repeating itself as a fire burns in front of the camp's central pavilion.

With a name that translates as 'resting place', this architect-designed standing camp is the main base for the Aboriginal-owned and -operated wukalina Walk, a four-day guided cultural hike along the Bay of Fires coastline. The wukalina Walk showcases the beauty of the famously colourful coastline of the Bay of Fires, but is just as focussed on culture, ranging from Creation stories and art to bush tucker.

Guests spend the first two nights of the walk in the camp, which consists of a central pavilion and individual sleeping pods dotted through the banksia scrub. Each charred-timber pod is accessed by a winch that ingeniously cranks open one of the walls to reveal a dome-shaped, blackwood-lined interior that replicates the shape of early Tasmanian Aboriginal shelters. Wallaby furs covering the beds provide additional warmth, and blackout curtains encourage slow waking. The central pavilion has a similar vaulted design, with a large deck that wraps around the central firepit, on which dinners of wallaby and muttonbird are cooked.

24

© ROB BURNETT IMAGES

🐢 THE PITCH

This award-winning standing camp and walk is as beautiful as the Bay of Fires beaches around it, and casts a spotlight on Tasmanian Aboriginal culture.

When: Sep-Apr
Amenities: tank water, compost toilets, firepit, all meals provided
Best accessed: by private shuttle
Cost: $2895 for the 4-day wukalina Walk
Contact: www.wukalinawalk.com.au

TAHUNE HUT
FRANKLIN-GORDON WILD RIVERS NATIONAL PARK

25

Few Tasmanian mountains dominate a scene quite like Frenchmans Cap. The 1446m (4744ft) mountain stands as a lofty sentinel over the Tasmanian Wilderness World Heritage Area.

At the foot of its white cliffs, on the shores of Lake Tahune, is one of Tasmania's newest and best mountain huts. The original Tahune Hut, built after WWII, was hewn from a single King Billy pine tree; this third iteration was opened in 2018 to replace the mould-ridden hut that had stood here for nearly 50 years.

Designed to fit the previous hut's footprint, the spacious new 26-bunk hut emerged like a butterfly – better, more beautiful, with rare wilderness comfort. The view is a given (step outside and you might crick your neck staring up at the mountain), but inside, powered by a hydro generator, there's lighting, heating and even USB ports. A drying room on entry contains heated racks for quick-drying of gear.

The insulation is exceptional, helping the hut retain a comfortable warmth, and the main living area is supplemented by a second room of bunks. Outside, a selection of tent platforms is sprinkled into the bush setting.

For most walkers, it's a two-day hike from the Lyell Hwy to Tahune Hut, stopping the first night at lower Vera Hut.

🧭 THE PITCH

One of Tasmania's newest wilderness huts, high on the slopes of the dramatic and imposing Frenchmans Cap, with a few unusual bush comforts.

When: Oct–Apr
Amenities: heating and lighting, USB ports, compost toilets, untreated water
Best accessed: by hiking (only)
Cost: free
Contact: https://parks.tas.gov.au

© ANDREW BAIN

KOMBI KAMP
HUON VALLEY

Few cultural icons embody the freedom of travel quite like a Kombi, and Frida wears her status well. This 1975 cyan-blue Kombi Transporter is parked beside a lovely stream running through a Huon Valley farm, within easy reach of wineries, cideries, orchards and fine restaurants.

The van is fitted out with a double bed, and on a fine night, you can throw open Frida's tailgate, pop something soothing on the sound system, and fall asleep to views of the stars.

The couples-only Kombi, which can be decorated in your choice of three themes, is parked among three rustic huts – a compost toilet, an open-air shower and a kitchen containing a gas stove, pot belly stove and all pots, pans and utensils are available. Or, you might prefer to cook on the firepit. Herbs can be picked direct from the garden, and there are farm-fresh eggs among the breakfast provisions.

It's the kind of place to check into to check out from the hustle and hassles of life. Instead, chill a wine in the eddy of the creek, take a nude swim in one of its inviting waterholes, or just sit quietly and you might spy a platypus paddling past.

THE PITCH

Live the nostalgic Kombi dream just a short drive from Hobart, sharing a sprawling farm with its Clydesdale horses and platypuses.

When: Sep-May
Amenities: gas stove, pot belly stove, hot shower, compost toilet, fridge, USB ports, breakfast provisions
Best accessed: by car
Cost: from $215 for 2 nights
Contact: www.airbnb.com

VICTORIA

You can camp by sand dunes in northwest Victoria, or by a snow-dusted mountain hut in the northeast. Grab provisions from gourmet towns, and fall asleep listening to waves pounding rugged shores.

Best time: Sep-Apr (camping/glamping), May-Aug (cabins and mountain huts)
Best national parks for camping: Alpine National Park, Wilsons Promontory National Park, Croajingolong National Park
Best camping trails: Grampians Peaks Trail, Falls to Hotham Alpine Crossing, The Wilderness Coast
National parks pass required: No
Useful contacts: www.parks.vic.gov.au; www.wildlifevictoria.org.au

As you travel Victoria's varied landscapes, from the mighty Murray River to the powerful Southern Ocean, you'll traverse many different traditional Aboriginal nations. The changes between each region are acute: notice the trees, wildlife, rocks, soil, and cloud cover above – on closer inspection Victoria is not a singular state but a collection of regions, each with a different character. Get to know the land you're on, and delve into its history for a rich connection.

For hardy campers, there are nature-based campsites that can only be reached by foot, canoe or bike. You need to take all your own supplies and carry out all your rubbish; and be prepared for any weather (yes, it can even snow in December in the mountains!), but you'll be rewarded with starry skies, abundant wildlife, and privacy.

There are also plenty of sites at some of Victoria's best locations where you can also bring your own car, trailer or campervan for a camping trip with all the bells and whistles. Some campgrounds require booking well in advance in the school holidays, or you may have to enter a ballot to get one of the coveted sites. You'll have more options if you visit outside peak times.

For lovers of the great outdoors who require a little more comfort, luxurious off-grid cabins abound, with amenities as impressive as the views.

FREE CAMPING

You need to book ahead and pay a fee for the popular or better serviced campsites across Victoria, but a handful remain completely free, particularly in more remote spots like Alpine National Park.

SUPPLIES

Outdoors stores like Paddy Palin, Bogong and Macpac are mainly concentrated in Melbourne/Naarm, while Aussie Disposals is in many regional centres too.

SAFETY

Deadly snakes inhabit Victoria including the Eastern Brown snake and Tiger snake. Get snake aware. Keep abreast of bushfire danger: know how to stay informed if there is any chance of local fires (www.emergency.vic.gov.au).

BEST REGIONS

Alpine country

Camp among soaring eucalypts or by a crisp mountain stream as far from civilisation as you can get in Victoria. Then hike to a mountain peak for expansive views of forested ridges in every direction.

Road tripping the Great Ocean Road (left); giant trees at Larnook flower farm (below)

Surf Coast

Crashing waves, dramatic cliffs, bushwalking and wildlife from kangaroos to koalas – the Surf Coast has plenty for nature lovers, plus towns with gourmet cafes and lively beer gardens for a touch of the urban, should you need it.

Grampians (Gariwerd) region

Rocky escarpments chiselled into a landscape over millennia, with stunning waterfalls and timid wildlife like echidnas and wallabies. This region is sacred to its Aboriginal communities; you can learn more about it at Brambuk The National Park and Cultural Centre in the region's hub town of Halls Gap.

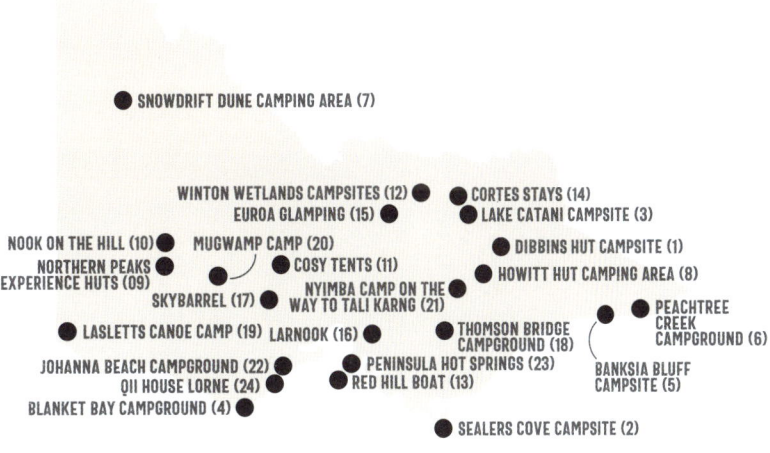

SNOWDRIFT DUNE CAMPING AREA (7)

WINTON WETLANDS CAMPSITES (12)
EUROA GLAMPING (15)
CORTES STAYS (14)
LAKE CATANI CAMPSITE (3)

NOOK ON THE HILL (10)
MUGWAMP CAMP (20)
NORTHERN PEAKS
EXPERIENCE HUTS (09)
COSY TENTS (11)
DIBBINS HUT CAMPSITE (1)
HOWITT HUT CAMPING AREA (8)

SKYBARREL (17)
NYIMBA CAMP ON THE
WAY TO TALI KARNG (21)
PEACHTREE
CREEK
CAMPGROUND (6)

LASLETTS CANOE CAMP (19)
LARNOOK (16)
THOMSON BRIDGE
CAMPGROUND (18)

JOHANNA BEACH CAMPGROUND (22)
PENINSULA HOT SPRINGS (23)
BANKSIA BLUFF
CAMPSITE (5)
OII HOUSE LORNE (24)
RED HILL BOAT (13)

BLANKET BAY CAMPGROUND (4)

SEALERS COVE CAMPSITE (2)

DIBBINS HUT CAMPSITE
ALPINE NATIONAL PARK

About halfway between Falls Creek and Mt Hotham in Victoria's High Plains, you'll find the Dibbins Hut campsite by the Cobungra River. Built out of felled logs and with a corrugated iron roof, the hut may look historic, but this is in fact a 40-year-old replica of the original 1920s cabin at the same site. Next to the hut is a firepit with a cooking plate and inside there is usually some firewood (plus an axe). With a picnic table to boot, it's a popular pit stop for day walkers as well as overnight campers.

Sleeping in the hut is only possible in an emergency but it's fine to shelter there from the rain – or sun. There is a designated camping area about 300m (984ft) downstream on the other side of the Falls to Hotham Alpine Crossing track. Numbered tent platforms make setting up camp that little bit easier (and they help protect the natural environment), but there are only five designated sites, so book ahead. There's another firepit at the campground, plus a pit toilet nearby. The best time to visit is from spring to late autumn, but don't rule out winter: with the right gear and a responsible attitude, staying out here after a dusting of white powder is especially magical.

🔖 THE PITCH

Camp among snow gums by a mountain stream and make use of this traditional mountain shelter for meal prep or huddling around a pot-belly stove in winter.

When: Sep-May
Amenities: 5 platforms for tents, fireplace, pit toilet
Best accessed: by foot
Cost: $15
Contact:
www.parks.vic.gov.au

01

02

SEALERS COVE CAMPSITE
WILSONS PROMONTORY NATIONAL PARK

Hike in to this idyllic campsite by a pristine, secluded beach, bracketed by silvery granite tors, on the east side of 'The Prom'. The walk here follows a well-maintained track that passes through lush ferny forests and along boardwalks over swampy marshes where streams run out to the ocean. This is not an easy stroll: be prepared for steep sections as well as changeable weather. You'll also need to plan your trip around the tide if you're coming from Telegraph Saddle, as there's an estuary to cross at low tide. Some campers stay at Sealers Cove as part of a multi-day Southern Circuit Walk, also stopping at Refuge Cove and Waterloo Bay. Both have similar basic campsites, but you're also likely to meet day-trippers coming by boat.

The Sealers Cove site has a pit toilet and access to creek water (which must be treated before drinking). Advanced booking is essential and there's a two-night maximum stay. Before setting off you'll need to swing by the Wilsons Prom visitor centre to check-in and get a proper briefing. The magic of this unspoiled otherworldly experience is dependent on every visitor respecting the site (which means, yes, carrying all your rubbish out).

🌐 THE PITCH

Only accessible by foot, this Wilsons Prom campsite takes you away from the hubbub of the main Tidal River campsite.

When: year-round
Amenities: unpowered site, no campfires, pit toilet, untreated creek water
Best accessed: by foot
Cost: $7.50
Contact: www.parks.vic.gov.au

LAKE CATANI CAMPSITE
MT BUFFALO NATIONAL PARK

03

There's a place for every type of camper at Lake Catani. Campervans and small caravans are catered for, and there are walk-in-only tent sites for those looking for more seclusion among the snow gums. Several sites cater for travellers with limited mobility with accessible toilets and showers (book ahead).

While there are excellent facilities here atop Mt Buffalo – gas BBQs, dishwashing sinks and sheltered picnic spots, plus flushing toilets and even showers – it is also remote. Campsites are unpowered, there's almost no phone reception, and it's a good hour's drive to the nearest shop for provisions. That said, once you're up here you won't want to leave. Mt Buffalo has multiple bushwalks to try: a leisurely stroll among the gums; a more challenging scramble among granite boulders with views of the plains below; or a heart-thumping hike down to a secluded waterfall and back. Lake Catani is also perfect for a paddle in a canoe or a cold-water plunge. Road cyclists love the challenge of climbing 1000m (3280ft) in just under 20km (12 miles) on the ride up from Porepunkah. There are also multiple rock climbing routes (book a guide with a local outfit). In winter you can snowshoe or toboggan at Cresta Valley or Dingo Dell.

THE PITCH

Idyllic in summer and a snowy wonderland in winter, this is an easy campsite stay with myriad outdoor activities on your doorstep.

When: Nov–Apr
Amenities: 49 sites, flushing toilets, showers, stone hut with tables and sink; BYO drinking water and firewood
Best accessed: by car
Cost: $28
Contact: www.parks.vic.gov.au

© GREG BRAVE | SHUTTERSTOCK

BLANKET BAY CAMPGROUND
GREAT OTWAY NATIONAL PARK

Whether you decide to walk in on the Great Ocean Walk (it's the second stop on the route from Apollo Bay) or arrive by car, Blanket Bay in the Great Otway National Park is an exceptional place to set up camp. Sites (some suitable for campervans) are dotted among tall manna gum trees. Most are sheltered from wind and sun, but a larger area for up to six 2- to 3-person tents is more open. Note that you need to keep food in airtight containers; native bush rats might gnaw your bags or into your tent if you don't.

This secluded campsite is perfect for coastal walks with big views. Head to Cape Otway lighthouse (10km/6 miles away) or stroll on the fern-laden Katabanut loop walk. Blanket Bay is also a prime rockpooling spot – pack a snorkel to explore it at high tide. For summer and long weekends like Easter, there's a ballot to secure one of the limited sites here. In the off-season, bookings are still required but there's less competition. Arrive in winter and you can look out for migrating whales passing just off-shore.

04

© SUPERJOSEPH | SHUTTERSTOCK

🔘 THE PITCH

Spot koalas in the eucalypts and listen to the waves at night at this tucked-away campground with 22 tent sites and direct access to the beach.

When: Sep-May
Amenities: unpowered tent sites, sheltered picnic tables, pit toilets; BYO water and firewood
Best accessed: by car or foot
Cost: $18
Contact: www.parks.vic.gov.au

BANKSIA BLUFF CAMPSITE
CAPE CONRAN COASTAL PARK

This popular yet remote campground has a prime location, near East Cape Beach with its pristine surf, multiple day walks (including through Yeerung Gorge on the traditional land of the Gunaikurnai people) and the lookout at Salmon Rocks where an Aboriginal shell midden stands.

There are bookable unpowered campsites at Banksia Bluff, as well as some sites available on a first-come basis (you'll have more luck outside the main school holiday periods). This is proper bush camping, albeit with flushing toilets. Expect only cold-water showers (using undrinkable bore water), no phone reception or wi-fi, and you'll need to BYO firewood and drinking water. It is possible to take campervans and camper trailers to many of the 135 sites here, but tents are more suitable for the smaller sites among the banksias and manna gums.

This is a region of significant Aboriginal cultural heritage, which is also recovering from devastating bushfires. Treat it with the reverence it deserves. Expect to see kangaroos, wombats, goannas – echidnas if you're lucky. Possums and kookaburras are rarer since the fires, but nature rebounds. May to October, look out for migrating whales off the coast.

◉ THE PITCH

Camp by the wild beaches of Cape Conran, now bursting with fresh green regrowth after the 2020 bushfires.

When: Sep-May
Amenities: cold-water showers and flushing toilets; BYO drinking water and firewood
Best accessed: by car
Cost: $18
Contact: www.parks.vic.gov.au

05

© KATACARIX | SHUTTERSTOCK

PEACHTREE CREEK CAMPGROUND
TAMBOON INLET, CROAJINGOLONG NATIONAL PARK

On the eastern side of Tamboon Inlet, Peachtree Creek campground is the epitome of remote tranquillity. A long way from a town, let alone a major city, your views of the stars are almost completely unaffected by light pollution here. This is also the perfect spot for birdwatching (listen out for kookaburras, currawongs or you may even spot the occasional white-bellied sea eagle), canoeing, kayaking and fishing. Anglers come for the flathead, salmon and perch in summer, and bream in winter.

Nearby, Point Hicks is of historic significance and the marine national park there is good for snorkelling (and diving), while Thurra River is best for wild swimming and sand dune climbing. However, at the time of research both spots were closed while the landscape recovers from the 2020 bushfires.

There's no need to book at Peachtree Creek – campsites (big enough for camper trailers, campervans and caravans) are available on a first-come basis – but you do need to pay. If you have a boat, canoe or kayak, you can push on to even more remote spots around the inlet. All sites are large enough for six people, are unpowered, and have basic pit toilets. You need to bring your own firewood and drinking water as none is available onsite.

06

🔴 THE PITCH

A remote and peaceful camp on the water's edge. With a boat or canoe you can cruise on to even more secluded spots across the inlet.

When: Sep-May
Amenities: 11 campsites, boat ramp, fireplaces, pit toilets; BYO water
Best accessed: by car (other nearby sites – and the coast – accessible by boat only)
Cost: $14.70
Contact: www.parks.vic.gov.au

SNOWDRIFT DUNE CAMPING AREA
WYPERFELD NATIONAL PARK

Snowdrift Dune is an utterly unexpected natural phenomenon. Between Big Desert country and flat mallee shrublands, you'll find large inland sand dunes appearing seemingly out of nowhere. This very basic campsite allows you to stay over and witness spectacular sunsets from atop the dunes. (BYO boogie board or sturdy piece of cardboard to toboggan down).

Covering 360,000 hectares, many visitors to the national park explore this largely flat, semi-arid landscape by 4WD, but campers can get to Snowdrift Dune by car, as long as it's dry.

This unique and fragile landscape was declared a national park in 1921. Within lies a series of (mostly dry) lake beds, a magnet for Australia's incredible native birdlife: 200 species have been spotted in the national park including Major Mitchell's pink cockatoos, multicoloured mulga parrots and local malleefowl – and, of course, there are plenty of kangaroos. This has been Wotjobaluk Country for tens of thousands of years; you may spot the occasional river red gum carved to make a canoe from wetter times when these lakes filled from an oversupply of water from the Wimmera River. Time your visit well and you'll see clusters of wildflowers carpeting the desert after rain.

07

THE PITCH

Victoria's third largest national park features inland sand dunes big enough to toboggan down. Spring brings an eruption of wildflowers across the flat mallee scrub.

When: Apr-Nov
Amenities: pit toilets, picnic table, firepits; BYO firewood and drinking water
Best accessed: by car
Cost: free
Contact: www.parks.vic.gov.au

© WRIGHT OUT THERE | SHUTTERSTOCK

HOWITT HUT CAMPING AREA
ALPINE NATIONAL PARK

Victoria's Snowy Mountain cattlemen hold an enduring place in Australia's national story, and this remote hut, originally built in the early 20th century but re-built over the last hundred years, is a part of that history. It's one of the oldest colonial structures in this section of the park. Archaeological work undertaken after the 2003 bushfires also confirmed the millennia of Aboriginal occupation here and the Alpine National Park authority recognises the deep and continuing connection the Taungurung and Gunaikurnai peoples have for Country.

Today visitors can either hike or drive up to camp overnight at Howitt Hut (no booking required). There's a relatively flat grassy area for tents, as well as a wooden picnic table, firepit and a pit toilet, but you'll need to supply your own firewood. If you're collecting water from the nearby stream, purify it. From here you can explore the alpine terrain, including the stunning waterfalls of Bryce Gorge on a 15km (9-mile) one-way day walk. Howitt Hut was protected from Victoria's recent bushfires by the Australian Army, who wrapped several alpine huts in a special foil-like material like a Christo artwork, protecting them from radiant heat and fire embers.

 **THE PITCH**

Camp among button grass and snow gums by a historic cattlemen's hut in the high plains. Regularly used as a base by hikers, it is also accessible by 4WD (or a 2WD with good clearance).

When: Sep-May
Amenities: pit toilet, untreated stream water
Best accessed: by foot or 4WD
Cost: free
Contact: www.parks.vic.gov.au

08

© TERRA INCOGNITA | ALAMY STOCK PHOTO

Mountain biking

Australia's southern states of Victoria and Tasmania have the nation's most developed and varied mountain biking scene, with bike parks and natural trails at every turn.

Blessed with a relatively mild climate, challenging topography, and environments that range from cool temperate rainforest gullies to boulder-strewn, granite highlands, Victoria stands out as the best place to try mountain biking in Australia. Its southern neighbour, Tasmania, is similarly blessed but perhaps demands a little more from its visitors.

Broadly, there are two different flavours of mountain biking: purpose-built, downhill-focused trails, usually crafted by trail builders around big hills such as Lake Mountain, Mt Buller and Mt Beauty; and more natural, cross-country trails, which range from tight and twisty trails at Forrest to long-distance A-to-B routes such as Victoria's Goldfields Track. For both, there are usually great camping options in the vicinity.

Your first decision is whether to bring your own bike or rent one. Mountain bike hire is often available near major trail networks, costing around $70-150 per day. But the standard of maintenance will vary. If you bring your own bike, you'll either need a vehicle or to carry all your kit on the bike, also known as bikepacking. Gear should include a helmet, comfortable clothing (padded shorts and no cotton close to the skin), compatible footwear, and a way to carry water, whether in bottles or a hydration pack.

WHERE TO RIDE

Selecting the perfect place to go mountain biking is the next step. In Victoria's northeast,

you'll find a huge variety of the most enjoyable trails in the country. Around Beechworth, local enthusiasts have connected the town with Yackandandah via a fantastic forest ride. Bright is another biking hotspot on the edge of the Australian Alps, offering challenging downhill tracks. Other mountain resorts include Mt Beauty, Falls Creek and Lake Mountain, with chairlift assistance available at Mt Buller bike park.

Out west, explore the Goldfields and the Otways on two wheels.

Creswick and Forrest both have good trail networks, plus longer routes on their doorstep. But close to Melbourne may offer the best yet: offbeat Warburton, bisected by the Yarra River, is becoming a brilliant hub for mountain biking.

Descending Gang Gangs Trail at Mt Buller (top); mountain town Bright is surrounded by trails; riding at Lake Mountain; the Goldfields Track (inset)

FIVE TO TRY

Mystic Bike Park, Bright
Swooping downhill trails through the pine plantations on the edge of this High Country town.

Yaugher Trailhead, Forrest
Twisty, sandy, cross-country trails set back from Victoria's Surf Coast in the Otways.

Warburton
Soon-to-be biking hub in the steep, misty forests of the Yarra Ranges, a train ride from Melbourne.

Mt Buller
This ski station becomes a bike park in the summer, known for its Alpine Epic trail and snow gums.

Creswick, the Goldfields
Out west in Victoria's Goldfields, Creswick offers some low-lying trails midway along the 200km Goldfields Track.

NORTHERN PEAKS EXPERIENCE HUTS
GRAMPIANS (GARIWERD) NATIONAL PARK

09

Traditionally, hikers haven't had many options for overnighting at the Grampians (Gariwerd) National Park beyond DIY camping. Since the opening of the 160km (100 miles) Grampians Peaks Trail, walkers can now traverse from north to south on a self-sufficient hike, staying at walk-in-only campgrounds with basic amenities. Alternatively, focus on a section of the park, depending on your preference for mountain ascents, cascading waterfalls, wildflower meadows or open forest.

You can now also stay in one of the eco-conscious huts in the Grampians for a more comfortable sleeping experience, without forgoing full immersion in this incredible, culturally rich landscape (Gariwerd – the Aboriginal name for the area – is vital to many local Aboriginal communities).

These lightweight, demountable wilderness huts each sleep up to four people in bunks. The open-plan spaces and slatted pergola system use less wood than a normal hut, making them lighter and kinder to the environment. The huts have small private decks and access to the shared cooking and toilet facilities. The Northern Peaks huts can only be accessed with licensed tour operators, who provide a variety of packages from guided hikes, abseiling and climbing, gear drops, and nocturnal tours revealing a world most of us are largely oblivious to unless shown by an expert guide.

THE PITCH

Low-impact modern huts designed for the ancient stone landscape of the Grampians. Only accessible on guided tours that offer varied levels of hiker support.

When: year-round
Amenities: 4-berth bunks, shared cooking and bathrooms
Best accessed: by foot
Cost: $1850 (tour dependent)
Contact: www.grampianspeaks.com.au/northern-huts

© ANDREW BAIN

NOOK ON THE HILL
HALLS GAP, GRAMPIANS (GARIWERD) NATIONAL PARK

10

Nook on The Hill sits proudly and sustainably on Djab Wurrung country. All the materials used to create this luxurious self-contained cabin have been salvaged and reclaimed: the 130-year-old bricks from a local bakery, oak from an Otways shearing shed, ironbark gates – this space is modern yet steeped in history.

Every remote tiny house has to make space and comfort trade-offs, but it's hard to see what The Nook is missing. You'll find polished concrete benchtops, a full-sized oven and hob, cleverly concealed dishwasher and handmade crockery that make cooking here a pleasure. And the tall roof and floor-to-ceiling windows not only give the cabin a sense of space, but also mean you can dine with views of the Grampians in the distance.

Soft linen sheets and cosy dressing gowns add to the touch of contemporary comfort in this ancient landscape. On a minimum two-night stay, you'll be greeted with a complimentary hamper of local goods including bread, spreads and wine from Pomonal Estate. Sip it while relaxing in the outdoor cast-iron bath under the stars, while tawny frogmouths catch moths and kangaroos and emus pass by.

THE PITCH

If the term tiny-mod-Scandi-chic existed it would describe this inspired owner-built dwelling at the foothills of the Grampians (Gariwerd) National Park.

When: year-round
Amenities: queen-size bed, cooks' kitchen, firepit, outdoor bath, native garden
Best accessed: by car
Cost: $650
Contact: www.nookonthehill.com.au

© HARLEY BROWN

COSY TENTS
DAYLESFORD, GOLDFIELDS

Accommodation around Daylesford and Hepburn once lent towards a certain heritage-chic aesthetic, a remnant of its century-old role as a place for immersion in its healing mineral waters. Recent years have seen a new type of stay emerge in this historic region, where a focus on sustainability, an appreciation for natural landscapes, and time spent connecting with each other around a primaeval camp fire, are the *raison d'être*.

Cosy Tents, in the village of Yandoit, a ten-minute drive north of Daylesford, is a prime example. It's an eco-friendly retreat with overnight options that range from Welsh-named bell tents to Scandi-style A-frame cabins, each catering to seekers of the simple life: good food, drink and company, in the elements – somewhere real (but with Bluetooth speakers). Subway tiles in the communal bathrooms, Hamptons outdoor chairs by the fire, and fairy lights draped in the camp kitchen bring a sophisticated domestic glamour to the bush glamping experience. This is a place where the natural world grows hotel attributes, and the result is a perfect dose of 'the bush' with shared community, and the ease and comfort of home (you can even add-on a Netflix package and binge some shows in bed).

THE PITCH

A-frame cabins and low-footprint canvas tents positioned halfway between Castlemaine and Victoria's spa capital, Daylesford.

When: year-round
Amenities: bell tents, A-frame cabins, fitted out camp kitchen, a communal BBQ area and firepits and shared bathrooms
Best accessed: by car
Cost: $295–395
Contact: www.cosytents.com.au

11

© FILEDIMAGE | SHUTTERSTOCK

WINTON WETLANDS CAMPSITES
BENALLA, NORTHEAST VICTORIA

A century-long restoration of a whole geography is taking place after the decommissioning of Lake Mokoan (a misguided manmade irrigation project) outside Benalla. So significant is the development of the Winton Wetlands, on Yorta Yorta Country, it has been declared a Wetland of Distinction by the Society of Wetland Scientists, the first such listing outside the USA. Pull up and you'll find a visitor centre and not-for-profit cafe, interpretative information signs, several campgrounds, picnic areas, toilet blocks, plus bush walk trails, cycling routes, and an art trail by 15 Yorta Yorta artists sharing stories of their living culture and their hope for the regeneration of this landscape.

You can get close to the kangaroos of the Bill Friday Swamp bushland campsite at sunset, join local stargazers gathering with telescopes at the Observation Pad near the Mokoan Hub & Cafe and listen for frogs and tawny frogmouths from your tent after dark. The multi-hued sunsets, all the more photogenic with the thousands of spindly gum trees taken by the ill-fated lake, have the look of something inspired by Salvador Dalí. Winton Wetlands employs a Koorie Cultural Officer to educate visitors on the Return to Country. Campers should be self-sufficient, bring plenty of insect repellent, and know what to do should any of their party be bitten by a snake.

12

© ROBERT WYATT | ALAMY STOCK PHOTO

🏕 THE PITCH

Camp by this reclaimed wetland with an art trail plus a huge variety of native flora and fauna — including mobs of kangaroos and 180 species of birds.

When: Mar–Dec
Amenities: 20 unpowered sites for tents and campervans, picnic shelters, communal firepit, pit toilet
Best accessed: by car
Cost: $20
Contact: www.winton wetlands.org.au

RED HILL BOAT
MORNINGTON PENINSULA

This dry-docked B&B – once a wartime rescue boat – keeps the home fires burning (in the outside firepit, that is) with a cosy mid-century feel. You'll find wood panelling, a Formica kitchen set, 1940s radio, papers and magazines, and vintage rattan outdoor chairs setting the scene. Below deck there are modernising touches including wi-fi and smart TV, electric blankets for winter, a sunken shower, and an outdoor bathtub large enough for two.

Red Hill is on the traditional lands of the Bunurong/Boon Wurrung people. Now a petite micro-region of the Mornington Peninsula, the area is an inland treasure trove of cellar doors, food stores, artisan producers, orchards and twice-hatted dining. Plus, there are millpond-still bay beaches and roaring surf beaches nearby.

To connect with nature, you can stroll through 17.5 hectares (43 acres) of remnant rainforest at Endeavour Fern Gully, or walk the 6.5km (4 mile) Red Hill Rail Trail to Merricks General Store. An afternoon soaking in the natural geothermal water at the Peninsula Hot Springs is another local highlight (see page 160).

THE PITCH

Stay on the Morning Peninsula at its most unusually nautical in this cosy boutique B&B (*boat* and breakfast) on a working farm kilometres from the sea.

When: year-round
Amenities: queen-size bed, kitchenette facilities, outside bath, BBQ, smart TV, Bluetooth sound system
Best accessed: by car
Cost: variable from $300 per night
Contact: www.redhillboat.au

13

© RED HILL BOAT

14

CORTES STAYS
OVENS VALLEY, NORTHEAST VICTORIA

In the mid-20th century, the Ovens Valley's post-colonial pastoralists, gold seekers and timber millers were joined by a wave of Italian migrants leaving a legacy of wine-making, olive groves and tobacco production. It's not surprising then that Cortes is located on a walnut farm, where accommodation includes a historic tobacco kiln restored and transformed into a house with a minimalist aesthetic as well as a modern off-grid cabin overlooking the Ovens River. The vibe is simple and rustic, but there are luxe touches like Aesop bathroom products, linen sheets, and, of course, plunger coffee (or coffee machine pods if you prefer).

Nearby you'll find vineyards, nut groves, berry and fruit farms drawing culinary tourists by bike, car and on foot to sample the valley's *la dolce vita*. The Great Alpine Road leads to outdoor adventures hub Bright in one

direction, and the produce-rich hamlets of Milawa, Stanley and Beechworth in the other. Any season is a good time to visit the Alpine National Park: spring brings wildflowers, autumn is best for leaf-peeping, and winter is ski season. Come evening, you can relax in the private outdoor bath, or take a dip in the alpine snow-melt flowing by your doorstep.

THE PITCH

Enjoy private bush views and river access at this small-house-design triumph in the epicurean hub of Victoria's Ovens Valley.

When: year-round
Amenities: queen-size bed, kitchen with gas cooker, outdoor bath, deck, BBQ; note no wi-fi
Best accessed: by car
Cost: $300-400 per night
Contact: www.cortes-stays. com.au

EUROA GLAMPING
STRATHBOGIE RANGES

15

This kid- and pet-friendly 50-acre glamping site is perched among the farmland, gumtrees and tussocky grasslands of northeast Victoria. Far from the coast, the air is often so still here you can hear a bird call a mile away, while richly coloured sunsets make way for staggeringly beautiful starry nights. The calm is also why Euroa is a popular parachuting drop zone. Back on Earth, the geodesic domes at Euroa Glamping are air-conditioned, with family-size fridge/freezers and a shared kitchen area, plus BBQs and seasonal campfires. Breakfasts and catering hampers are available on request, and staff are only too keen to ensure you enjoy your stay.

The Strathbogie Ranges are on Taungurung Country. With three under-the-radar national parks (Lake Eildon, Lower Goulburn and Warby-Ovens) within an hour's drive, there's plenty of opportunity to connect with the region's unique landscapes and native flora and fauna. Sunrises at the Euroa Glamping come with views of horses and cows in a neighbouring paddock, and a walking track around the site reveals rolling hills. Many know this as (Ned) 'Kelly country' and drives around the area, from the Strathbogies to Avenel and Glenrowan, are rich for post-colonial tales of injustice, displacement and daylight robbery.

 **THE PITCH**

Family-friendly glamping in an underrated region. See kangaroos jumping by at sunset then sit by your own fire pit and watch the dramatic night sky light up.

When: year-round
Amenities: six powered geo-dome tents, BBQs, air-con, wi-fi, campfires, shared kitchen and bathroom facilities
Best accessed: by car
Cost: from $250
Contact: www.euroa glamping.com.au

16

LARNOOK
DANDENONG RANGES

Tucked up by a charming refitted shed on a huge flower farm, Larnook offers tiny-dwelling chic for a nature-based break only an hour from Melbourne. Taken from the Boon Wurrung word for habitat, Larnook comprises two adjoining modified off-grid shipping containers, decorated with designer flair. The list of luxe additions is long – indoor gas-log fireplace, Nespresso coffee machine, Bose Bluetooth speakers, top-shelf bathroom products, plus a bottle of complimentary local wine and chocolate – and that's on top of the views from the deck. There's plenty of bushwalking to keep you connected to nature,

from an easy forest stroll to walks sufficiently steep to get the heart rate up. An hour-long walk through Sherbrooke Forest brings you to Grants Picnic Ground, famous for its crimson rosellas and rainbow lorikeets, while quaint mountain

villages like Tecoma, Kallista and Olinda have their own charms. You're also close to the Puffing Billy Railway. For gourmet touring, the Yarra Valley – with its wineries and hamlets like Healesville and Warburton – spread out to the northeast. But a stay at Larnook is likely to have you ignoring the outside world's day-trip offerings in blissful comfort.

💲 THE PITCH

A solo or couple's retreat in an off-grid hideaway with lashings of luxury and day-trip options galore.

When: year-round
Amenities: king-size bed, indoor and outdoor hot shower, domestic kitchen and supplies, and a firepit
Best accessed: by car
Cost: $790 for 2 nights
Contact: www.larnook.co

The Great Bathing Trail

A new touring trail connects many of Victoria's bathhouses, spas, hot and mineral springs, for those who love to soak or float in water, most with camping options nearby.

Think of spas and bathing in natural hot or mineral springs in Victoria and most people will imagine Daylesford, a couple of hours northwest of Melbourne. This Goldfields town is surrounded by mineral springs and is also a famously gay-friendly community. However, a new initiative is hoping to broaden the horizons of people who love to bathe by creating the Great Victoria Bathing Trail.

Presently, the trail extends from the east of Gippsland across the state to the west of Warrnambool, with Melbourne at its centre. And Melbourne is where most people will start: although there are few places to camp or stay overnight in a hut, it would be a shame to miss swimming in the ocean off St Kilda or Sandringham. Plans to introduce natural bathing pools to the Yarra River are also in motion.

WHERE TO BATHE

East of Melbourne, the Great Victorian Bathing Trail travels down the Mornington Peninsula to Peninsula Hot Springs and Alba Thermal Springs and Spa, where geothermal pools are set in landscaped gardens. Continuing east, there are hot springs on Phillip Island and around Traralgon in Gippsland, before you reach Metung Hot Springs near Lakes Entrance. At Metung Hot Springs, a drive of about four hours from Melbourne, steaming hot water is siphoned from an aquifer 500m below ground before being cooled to a comfortable temperature. Guests may stay under canvas in glamping-style tents by the water.

To the north of Metung, bathers can soak in an onsen at Mt Hotham. Returning to Melbourne, Daylesford and Hepburn are a short hop north. The region is noted for its mineral springs, intended for sipping as much as dipping. The springs are rich in minerals and each has its own taste: try Taradale or Glenlyon. Or go for a swim in Jubilee Lake. Adjoining Daylesford, Hepburn's Mineral Springs Reserve has a number of different springs where you can top up your water bottles. There are a number of free campgrounds a short drive from Daylesford.

Out west, the Great Victorian Bathing Trail follows the coast from Geelong to the sea baths at Lorne and onward to hot springs beyond the Twelve Apostles.

Tubs with a view at Metung Hot Springs (top); soaking in the surroundings at Peninsula Hot Springs (above and inset); Lake Daylesford (left) in Victoria's spa capital

FIVE TO TRY

Daylesford and Hepburn
Long Victoria's original spa playground, there are numerous mineral springs to sample here.

Mornington Peninsula
Mornington is a hotbed of thermal activity at Peninsula Hot Springs, plus sea bathing.

Geelong
There's sea bathing at Geelong's Eastern Beach plus mineral springs at a wellness centre.

Melbourne
Dip into sea bathing off St Kilda and Sandringham and mooted natural pools on the Yarra.

Gippsland
Metung Hot Springs, Victoria's new geothermal bathing experience, offers waterside glamping.

17

SKYBARREL
GOLDFIELDS

The latest addition to an enigmatic collection of luxury and eclectic accommodation in the Ballarat region, Skybarrel has had plenty of publicity since opening in 2022. Superlatives and exclamation marks abound in the visitor reviews, and you can see why. This luxury stay, designed by local architect Robin Larsen, has been described as UFO-like. Inside, the cabin has a minimalist, industrial vibe but with soft touches like fluffy white towels, muted-tone linen bedding, and a cowhide rug.

Perched on land to the side of the now-extinct 745m (2444ft) high volcano at Mt Buninyong which is part of the traditional lands of the Wathaurong nation, Skybarrel affords guests sweeping views of native bush and the plains below. Upmarket touches include a large-screen TV in the salon and another in the downstairs bedroom, but with floor-to-ceiling windows

🫖 THE PITCH

An industrial chic chalet suite inside a giant barrel by Mt Buninyong, with architectural finishes, an indoor firebox, wooden bath, and sheer vertical audacity.

upstairs the view of the passing cloudscapes and the nighttime star show may be more entertaining. The owners are also responsible for the much-loved Clifftop Hepburn (also on Wathaurong Country), with a similar industrial style (rust exterior, glass floors, and a shipping container pool) and with a balcony to breathe in the fresh country air.

When: year-round
Amenities: double bed, ensuite bathroom, two large-screen TVs, and petite kitchen
Best accessed: by car
Cost: $800 for 2 nights
Contact: www.clifftopathepburn.com.au/skybarrel-villas

© SKYBARREL

18

THOMSON BRIDGE CAMPGROUND
WALHALLA, GIPPSLAND

Gippsland's rolling hills of dairy country, pocketed with rainforest-like temperate bushland, are a road tripper's dream. Off the main highways, you'll find scenic drives including the road to the historic gold mining town of Walhalla. Continuing the picturesque – and historic – theme, just a 10-minute drive (or a short steam-train ride) away is Thomson Bridge, where you'll find the recently upgraded Thomson Bridge Campground. At the site you'll find a secluded clearing tucked into lush bushland. Facilities include five unpowered sites available on a first-come basis, basic toilets, a fire pit, and sheltered picnic tables. BYO water

or purify water from the river.

There's a host of outdoor activities, from riding the Walhalla Goldfields Railway to tackling the Thomson River Canoe Trail. Walk to Poverty Point Bridge to take in views of the river, cycle the

Walhalla Goldfields Rail Trail to Erica, or ride the 30km (19 miles) of track at Erica Mountain Bike Park. The Walhalla to Thomson Bridge Station steam train runs at night in winter, and Wednesdays and weekends all year.

🐾 THE PITCH

Gippsland's experience of the Gold Rush left a historic ghost town in this valley with myriad outdoor activities.

When: all year
Amenities: firepits, sheltered picnic tables, toilets
Best accessed: by foot, bike or car
Cost: free
Contact: www.explore outdoors.vic.gov.au/activities/camping/thomson-bridge-campground

19

LASLETTS CANOE CAMP
LOWER GLENELG NATIONAL PARK

On the traditional lands of the Gunditjmara people, and almost equidistant between Melbourne and Adelaide, the Glenelg River meanders through limestone gorges and forested riversides before it spills out into the Southern Ocean. The best way to see it is by kayak or canoe, although it is possible to explore the river and surrounding national park by foot.

There are a number of campsites along the river to overnight at, some of which were upgraded in 2023. Lasletts Canoe Camp is one of the larger sites, with plenty of flat grassy land to pitch a tent. There are also pit toilets, a rain-water tank, firepits, picnic tables and a jetty. It's a three- to four-kilometre paddle downstream to the Princess Margaret Rose Caves, an underground limestone cave system and one of the main tourist draws in the region. The Glenelg River Canoe Trail starts at Dartmoor and ends in the township of Nelson, although less experienced canoeists might want to skip the first section and begin their tour at Moleskin. You can hire canoes and gear in Dartmoor, Winnap or Nelson and most outfits will arrange drop-off and pick-up points along the river depending on how much time you have.

⬤ THE PITCH

Pitch a tent with a view of the mottled limestone cliffs reflected in the Glenelg River on this canoe trail near the Victoria and South Australia border.

When: Mar–Nov
Amenities: fireplace, jetty, picnic table, pit toilet, seasonal tank water for boiling
Best accessed: by canoe or foot
Cost: from $6
Contact: www.parks.vic.gov.au

© ANDREW BAIN

MUGWAMP CAMP
MT COLE STATE FOREST, PYRENEES REGION

Located on the Mt Cole plateau, on the traditional land of the Eastern Maar peoples, is a historic hut, originally built for forest workers supplying building materials and firewood for nearby goldfield towns.   These days, the stone and corrugated iron structure is not used for accommodation (except for shelter from rare snow, rain or sunshine) but the campground around it welcomes overnight visitors. Designated fire spots, basic toilets and a picnic table make this a comfortable campsite, although it's best visited outside the height of summer.

One of the best reasons to stay at Mugwamp Camp, besides the excellent facilities, is to climb the final mile up Mt Buangor to watch the sun rise or set while overlooking the plains below. Mt Cole itself is home to a unique blood-red grevillea plant, which depends on the region's altitude and granitic soils to bloom. You'll also spot kangaroos, wallabies, echidnas and koalas, plus 130 different birds including kookaburras, not to mention nocturnal animals like possums.

The campsite, surrounded by gum trees and fern shrubs, has limited parking, or you can walk in on the Mt Buangor walking track.

🧭 THE PITCH

Camp among eucalypt trees, next to a former forest workers' hut, and climb to the top of Mt Buangor to watch the sun rise or set.

When: Mar–Nov
Amenities: tent camping, BYO water, uncovered tables, basic toilets, wood BBQ and firepits
Best accessed: by car or foot
Cost: free
Contact: www.visit pyrenees.com.au/seeand do/camping-in-mount-cole

20

NYIMBA CAMP ON THE WAY TO TALI KARNG
ALPINE NATIONAL PARK

21

Tali Karng is not an easy place to get to, which means you're more than likely to have this secret lake, nestled away in Victoria's High Plains, all to yourself. You're on Gunaikurnai Country, and the original custodians of the region ask visitors to stay away from the lake after dark. (According to Parks Victoria, there is no opposition to non-indigenous people visiting Tali Karng, if they treat the area with respect.) The wilderness Nyimba Camp is the closest place to stay as part of an overnight walk. From there, it's a hard five-hour walk down (an 800m/2625ft descent) and back up. Most campers leave their gear at Nyimba and travel light for this section.

The alpine campsite is on the southern part of the Wellington Plains, in a relatively flat clearing between the twisted white trunks of snow gums. The closest spot to collect water (which will need to be filtered) is Nigothoruk Creek, 650m (2132ft) away on the track to Millers Hut. Downstream you'll also find the picturesque Nigothoruk Falls. Note: there's no toilet here due to previous bushfires, so bring a trowel and camp responsibly. Also nearby is the pinnacle of Mt Wellington (at 1634m/5361ft), a 4.5km (3-mile) walk along the Wellington Plains walking track, with expansive views over blue-hued mountains in every direction. It's also accessible by 4WD on the Mt Wellington track.

🧭 THE PITCH

A remote mountainside campsite on the Wellington Plains walking track. Overnight here to visit lake Tali Karng, Victoria's highest natural stream-fed lake, and a sacred site for the Gunaikurnai people.

When: Sep-May
Amenities: nearby creek, no toilets (BYO trowel), firepit
Best accessed: on foot
Cost: free
Contact: www.parks.vic.gov.au

JOHANNA BEACH CAMPGROUND
GREAT OTWAY NATIONAL PARK

Johanna is an expansive and unpatrolled beach between the Cape Otway Lighthouse and the world-famous Twelve Apostles on Victoria's southwest coast on Gadubanud Country. This is a popular place for surfing, surf fishing, and long ambles along the stunning shoreline. It's a nesting area for the endangered hooded plover, so visitors are expected to stay below the high-tide line and out of the dunes, and to keep dogs on leads. To better imagine this ancient forested landscape before colonisation, pop by the pocket of lush rainforest at Melba Gully, just off the Great Ocean Road.

While the campsite here is basic, it's hugely popular for its location and stunning views. From May to October, humpback whales pass on their annual migration from Antarctica. Some campers walk in as part of the Great Ocean Walk, but many arrive by car or campervan. There are 25 unpowered sites available to book; amenities are limited, and you'll need to carry your own drinking water. Note also: fires are not allowed at any time, so you'll need a camp stove.

⊜ THE PITCH

On Victoria's dramatic Surf Coast, this simple, secluded campsite is nestled in a grassy clearing among the sand dunes.

When: Sep-Jun
Amenities: 25 camp spots, BYO water, compost toilets, no shower
Best accessed: by car
Cost: $16
Contact: www.parks.vic.gov.au

PENINSULA HOT SPRINGS
MORNINGTON PENINSULA

Located in the heart of the Mornington Peninsula, these geothermal mineral springs offer visitors a huge variety of spa experiences, all in a beautifully landscaped gully of Australian coastal bush. Spa options start from a soak in communal Japanese-style spring pools right up to full private bathing. There's also an ice cave and underground sauna, massage treatments, reflexology and a hamam steam room onsite. Since opening in 2005 the visionary owners also added a cafe, wellness centre, an events amphitheatre, plus 'Sunday sessions' with local musicians, and even an artist-in-residence programme.

Don't want to leave? You can sleep over in luxury glamping tents replete with king-size beds, ensuite bathrooms with rain showers, heated floors, and private decks where you can sip a glass of wine while listening to the call of native birds and frogs, and the burbling of the nearby springs. The glamping here is not cheap, even for an entry level stay (check the website for midweek discounts). For those with plenty of coin, there is a raft of additions to make this a once-in-a-lifetime experience you'll be reminiscing about for years.

THE PITCH

There's glamping and then there's wellness glamping with in-room dining, yoga, overnight hot springs bathing, premium breakfast, and spa treatments to round it all off.

When: year-round
Amenities: 10 glamping tents
Best accessed: by car
Cost: $670
Contact: www.peninsulahotsprings.com

23

© NATALIA VOSTR KOVA | SHUTTERSTOCK

24

QII HOUSE LORNE
GREAT OTWAY NATIONAL PARK

Deep in the Otway Ranges, 550m (1804ft) above sea level, where snow falls in winter, this modern and quirky eco retreat was recently voted one of Australia's best self-contained places to stay. The location is amazing, with ancient rainforest at your window and the lush Erskine Falls accessible by foot, plus Lorne's shops, restaurants and popular surf beach a 20-minute drive– but another world – away.

Reconnecting with nature is very much the guiding principle behind this cabin, where visitors are invited to practise yoga and meditation, or simply sit and listen to the kookaburras in the forest. The large timber house, decked in 60s-style furnishings, sleeps eight, which makes this three-night-minimum retreat a favourite for larger families, groups of friends and even colleagues looking to recharge. There are also several tiny houses for two dotted around.

The owner renovated the property in keeping with its eye-catching modernist architecture – the large 'fishbowl' windows in particular – but added contemporary touches like a private deck with a custom-built outdoor bath for two. The tiny houses are designed with *shinrin-yoku* (forest bathing) in mind. A Japanese-style garden also stands out in the natural bushland.

  THE PITCH

Get forest bathing on your weekend-long stay at this Japanese-style retreat in the Otway Ranges where nature, meditation, and modernist architecture commingle.

When: year-round
Amenities: kitchen, lounge and deck plus several tiny house cabins on-site
Best accessed: by car
Cost: $299 to $1000
Contact: www.qiihouselorne.com.au

© GARETH JAGO | PROPELD

WESTERN AUSTRALIA

Red deserts, turquoise bays and Kimberley wilderness converge in Australia's largest state, where long journeys to remote campsites are just as memorable as the destination.

Best time: May–Oct (Kimberley, outback), year-round (South West)
Best national parks for camping: Cape Le Grand/Mandoowernup National Park, Cape Range National Park, Karijini National Park, Leeuwin-Naturaliste National Park, Purnululu National Park
Best camping trails: Bibbulmun Track, Cape to Cape Track
National parks pass required: Vehicle pass required for most parks
Useful contacts: www.westernaustralia.com; https://exploreparks.dbca.wa.gov.au

Ten times the size of the United Kingdom, Western Australia really is huge. And with 80 per cent of its 2.8 million residents based in Perth, that leaves a whole lotta wilderness to explore.

From the ancient, rocky wilderness of the Kimberley to the crystalline beaches of the South West, there's a never-ending supply of epic landscapes layered with culture to explore in Western Australia. With much of its iconic nature located miles from the closest town, camping is the best way to experience outdoors WA, especially during the golden hours of sunrise and sunset when the vibrancy levels are dialled up and create pure magic. While the wet season makes travelling north of Broome tricky, it's a great time to hit the southern beaches.

FREE CAMPING

Camping fees apply in most national parks, with good free options in Stokes National Park and Peak Charles National Park, both near Esperance. Some local councils offer free sites; Betty's Beach, 50km (31 miles) east of Albany, is a standout. In the Goldfields, it's free to camp at Karalee Rocks on the National Trust-listed Golden Pipeline Heritage Trail.

SUPPLIES

Perth has the widest selection of camping and outdoors stores; Tentworld (www.tentworld.com.au) in Midland is one of the chain's largest stores. If you're heading south from Kununurra, the Bushcamp Surplus Store (www.facebook.com/BushcampSurplus) is the place to grab last-minute supplies.

SAFETY

The Kimberley is croc country, and also experiences stinger season (Nov–May) from the Broome region northward; always heed warnings. Keep an eye on shark activity with the SharkSmart WA app, and on bushfire warnings via www.facebook.com/dfeswa. Driving in the Kimberley is dicey in the wet season (Nov–Apr), when roads can flood. With these long drives, travelling with ample water and fuel is a must.

BEST REGIONS

Kimberley

Surreal rock formations, thundering waterfalls, dramatic gorges and Aboriginal culture are all highlights of visiting the vast wilderness occupying the northern end of the state. Load up a 4WD with all the essentials for a remote camping adventure, and get amongst it.

Safari style at Tanja Lagoon Camp (left);
a quokka on Rottnest Island (below)

South West

Wineries and old-growth forests
fringe some of Australia's most
beautiful beaches in the state's
southwestern corner, with plenty
of national parks, hidden bays,
and coastal trails to explore.
Between August and November,
more than 8000 wildflower
species burst into bloom.

The Coral Coast

The outback meets the sparkling
Indian Ocean on this 1100km/684-
mile stretch of coastline north of
Perth/Boorloo, with blockbuster
seaside camping at two World
Heritage Sites: Shark Bay and the
Ningaloo Coast.

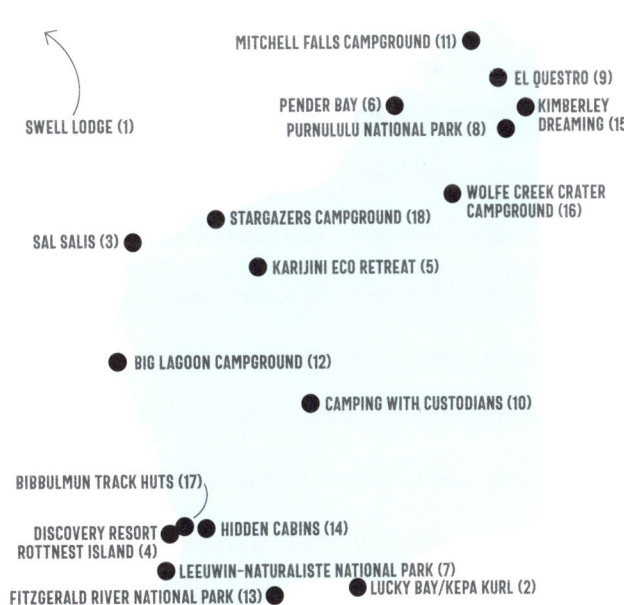

MITCHELL FALLS CAMPGROUND (11)

EL QUESTRO (9)

PENDER BAY (6)
PURNULULU NATIONAL PARK (8)

KIMBERLEY
DREAMING (15)

SWELL LODGE (1)

WOLFE CREEK CRATER
CAMPGROUND (16)

STARGAZERS CAMPGROUND (18)

SAL SALIS (3)

KARIJINI ECO RETREAT (5)

BIG LAGOON CAMPGROUND (12)

CAMPING WITH CUSTODIANS (10)

BIBBULMUN TRACK HUTS (17)

DISCOVERY RESORT
ROTTNEST ISLAND (4)

HIDDEN CABINS (14)

LEEUWIN–NATURALISTE NATIONAL PARK (7)

FITZGERALD RIVER NATIONAL PARK (13)

LUCKY BAY/KEPA KURL (2)

SWELL LODGE
CHRISTMAS ISLAND NATIONAL PARK, CHRISTMAS ISLAND

01

When the blazing sunset fades to inky black and your private chef has packed up and left for the evening, the real magic begins at Swell Lodge. The only accommodation in Christmas Island National Park, this pair of deluxe, privately situated cliff-top glamping tents on the island's remote west coast offer the rare chance to experience this jungle wilderness at night, when its famous crabs are most active. While Christmas Island is best known for the annual migration of these vibrant red crustaceans, which can be seen year-round, the mostly nocturnal robber crabs are a similarly incredible sight to behold as they emerge to forage on the forest floor. Weighing up to 4kg, it's impossible to miss them.

Meals and daily activities are included at Swell Lodge, which was designed to create the smallest possible footprint on this primeval landscape. Power comes from the sun, air-con is provided by the sea breeze, and the evening entertainment is a thrilling choice between a self-guided spotlighting safari and stretching out on your deck, using the stargazing app on your in-tent iPad to decode the sparkling sky. From your front door it's an easy, 4km (2.4-mile) return walk to the beautiful Hughs Dale Waterfall, where you're likely to spot endemic sky-blue crabs.

 **THE PITCH**

Wake to tropicbirds dancing above the Indian Ocean just metres from your bed at the only place to spend the night in the national park covering most of remote Christmas Island.

When: year-round
Amenities: drinking water, toilets, electricity, wi-fi
Best accessed: by plane from Perth, then car
Cost: $5385 for 2pp for 3 nights
Contact: https://swelllodge.com

© ULI KUNZ

LUCKY BAY/KEPA KURL
CAPE LE GRAND NATIONAL PARK/MANDOOWERNUP

Between its blinding white sands and electric blue water, it's easy to see why this protected, crystal-clear bay near Esperance in the state's southwest (700km/435 miles from Perth) regularly tops world's best beaches lists. Framed by scenic granite outcrops linked by bushwalking trails, Lucky Bay is also popular with kangaroos, which can often be seen grazing by the beach.

At the western end of the bay, a national parks-run campground offers the chance to spend a few days (up to 14 if you can't bear to leave) in this spectacularly beautiful slice of nature. Advance planning is key here – its 56 reverse-in spots, buffered from the wind by coastal scrub, book out fast. And there are seasonal weather patterns to consider: December to April can be very hot, while nights can be cold from June to September, and storms are common between May and October. Whales cruise by from July to October, and consistently windy days between November to early March make the shallow bay popular with kitesurfers. Tame kangaroos are most commonly sighted on and around the beach at dawn and dusk year-round; resist the urge to touch or feed them.

02

© ANEK.SOOWANNAPHOOM | SHUTTERSTOCK

🌿 THE PITCH

Why just spend the day at one of Australia's best beaches when you can stay in this outrageously beautiful coastal landscape for nights on end?

_____

When: year-round
Amenities: untreated water, toilets, showers, food preparation shelters, barbecues, waste facilities
Best accessed: by car
Cost: $15
Contact: https://exploreparks.dbca.wa.gov.au

SAL SALIS
CAPE RANGE NATIONAL PARK

The tropical blues of Ningaloo Marine Park meet the arid coastal plain of Cape Range in a spectacular clash of contrasting scenery in Cape Range National Park, southwest of Exmouth. For self-sufficient campers, the 10 basic, 2WD-accessible national parks-run campgrounds along the waterfront offer an opportunity to explore this harsh but beautiful landscape on the cheap. For everyone else, there's Sal Salis. Discreetly tucked in the dunes just steps from the reef, this iconic glamping stay offers a more luxurious immersion in this special place. Rise at dawn to snorkel over fields of neon-blue staghorn coral before a sumptuous breakfast on the communal outdoor dining deck. Then, take your pick from a range of guided activities, from hikes into ancient canyons to snorkelling sessions at the best spots along the coast. Or you might decide to head offsite with a local operator to experience the magic of swimming (responsibly) with whale sharks between mid-March and the end of September. Back at camp, evening canapes kick off a three-course al fresco dinner, with the open bar tempting a tipple to be savoured as the ocean shimmers in the glow of a million stars.

03

🌊 THE PITCH

The vibrant coral gardens of the World Heritage-listed Ningaloo Reef lie just steps from your luxury glamping tent at this exclusive coastal camp.

When: mid-Mar–mid-Nov
Amenities: drinking water, private bathroom, electricity, restaurant, open bar
Best accessed: by car
Cost: from $1590, 2-night minimum
Contact: www.salsalis. com.au

© BEN FUNNEKOTTER

04

DISCOVERY RESORT ROTTNEST ISLAND
ROTTNEST ISLAND/WADJEMUP

Crystal-clear water, snow-white sandy beaches, and one of Australia's cutest animals – the photogenic quokka – await on 'Rotto', a popular holiday retreat just a 25-minute ferry ride from Fremantle. Forming part of the traditional Country of the Whadjuk Noongar people, the island's Indigenous name, meaning Place of Spirits, is the first hint of its rich history for Aboriginal people across Western Australia. Connect to the historical, cultural and spiritual significance of the island on an Aboriginal-guided tour with Go Cultural, and a visit to the Wadjemup Museum. Then jump on a bicycle (or the hop-on, hop-off

bus) to explore the car-free island.

BYO tent to bed down at an unpowered site at central campground Stay Rottnest Island, or opt for a glamping tent at the eco-conscious Discovery Resort Rottnest Island, next door. The

Deluxe Oceanside tent is our pick. Fall asleep watching the stars twinkle above the Indian Ocean, and wake to waves lapping the shores of Pinky Beach. While endemic quokkas make for great photos, try not to crowd them.

 THE PITCH

Cast away from Fremantle to glamp alongside one of Australia's most Instagrammable marsupials and allow 'Rotto' to reveal even more of its magic after the day trippers have departed.

When: year-round
Amenities: drinking water, toilets, showers, electricity, barbecues, waste facilities, wi-fi at the beach club
Best accessed: by ferry
Cost: glamping from $299
Contact: www. discoveryholidayparks.com.au

KARIJINI ECO RETREAT
KARIJINI NATIONAL PARK

Take an epic journey back in time to Karijini National Park. Out here, on the traditional lands of the Banyjima, Kurrama and Innawonga people of the Pilbara region, some 1400km (870 miles) northeast of Perth, erosion has carved spectacular gorges out of the red-rock landscape. After a long day's drive from Exmouth or Broome (possible by 2WD), you'll want to spend a few days in this surreal landscape. If you're self-sufficient, the national parks-run Dales Campground might suit your needs. For more comfort, the Aboriginal-run Karijini Eco Retreat is the place to be, with unpowered campsites, eco cabins and a range of glamping tents complemented by an outback restaurant.

It's only a short walk from the retreat to Joffre Gorge, one of the park's highlights with its scenic rocky amphitheatre, waterfall and refreshing plunge pool. A short drive away, Weano Gorge's idyllic Handrail Pool is well worth the scramble, as is navigating the 'spider walk' to nearby bright-green Kermits Pool. And there are plenty of other ancient gorges to explore, as well as crystal-clear freshwater pools to cool off in, before you bed down under a cosmic canopy. For a closer look through a telescope, sign up for the retreat's Night Sky Tour.

THE PITCH

Embrace the middle-of-nowhere beauty of Western Australia's largest national park, where an Aboriginal-owned glamping outfit offers a welcome dose of comfort after a day exploring local gorges.

When: Apr–Oct
Amenities: toilets, showers, electricity, restaurant
Best accessed: by car
Cost: eco tent from $240
Contact details: www. karijiniecoretreat.com.au

05

© AGENT WOLF | SHUTTERSTOCK

06

PENDER BAY
DAMPIER PENINSULA

The teeming turquoise waters of the Indian Ocean meet the rich red dirt of the Kimberley in an explosion of colour on the Dampier Peninsula, north of Broome. Here juicy mud crabs hide in lush mangrove forests, prized fish tempt keen anglers, and Aboriginal culture threads through the land. This remote landscape has sustained the Bardi Jawi people for centuries, and today their descendants run a number of good-quality campgrounds with a range of facilities including Djarindjin Campground, Goombaragin Eco Retreat and the Lombadina community campground.

But there's something very special about camping on the fiery cliffs of the Dampier, particularly at Pender Bay Escape, a simple, Aboriginal-run campground with a handful of sites overlooking the sapphire sweep of Pender Bay. It's a long, dusty 4WD-only drive to get here from Broome, but with views from your tent like this – especially when the setting sun dances on the pindan cliffs, and one of the world's largest populations of migrating humpback whales visits in August – it's worth it.

Further south on the Dampier, James Price Point is a similarly scenic beach camping spot with no facilities, and it won't cost you a cent to pitch up here.

THE PITCH

The pindan (red-soil) cliffs of this rugged peninsula stretching north of Broome are dotted with camping areas, with one cliff-top stunner to rule them all.

When: May–Oct
Amenities: toilets, showers
Best accessed: by car (4WD)
Cost: $20
Contact details:
www.facebook.com/
penderbayescape

LEEUWIN-NATURALISTE NATIONAL PARK
SOUTH WEST

Magnificent limestone caves, dreamy turquoise bays and towering native forests are just some of the natural draws at Western Australia's southwestern tip. Only a three-hour drive south of Perth, the long, skinny Leeuwin-Naturaliste National Park, stretching between its namesake capes, is one of the state's more accessible wilderness areas, linked from top to bottom by the 125km (78-mile) Cape to Cape Track. While the cooler months are great for hiking, the idyllic rock pools along the coast offer a series of refreshing diversions on warmer days as dolphins and surfers ride the waves beyond.

There are five campgrounds to choose from in the national park. Feel the sand between your toes – and a hot shower warm your bones at the end of the day – at privately-managed Hamelin Bay, where stingrays cruise the shallows. Or go bush and pitch up alongside kangaroos and quendas (southern brown bandicoots) at the large, sheltered Conto Campground, conveniently close to Conto Beach. Medium-sized Jarrahdene Campground also takes advance bookings; or you can try your luck rolling into the smaller non-bookable Point Road Campground or Boranup Campground (bring cash). All options offer an immersion in nature within easy reach of regional town centres – and plenty of wineries.

07

THE PITCH

Just a stone's throw from the premium wineries of the Margaret River region lies a protected stretch of coastline full of drama and wonder. Hike, swim, surf and camp amid its natural splendour.

When: year-round
Amenities: toilets, plus barbecues, picnic tables and fire rings at Conto Campground
Best accessed: by car
Cost: from $11
Contact: https://exploreparks.dbca.wa.gov.au

© MARIE HENSON | SHUTTERSTOCK

PURNULULU NATIONAL PARK
KIMBERLEY

08

Rising up from the spinifex grasslands like giant beehives, the banded domes of the Bungle Bungle Range largely remained the secret of its Traditional Custodians, the Jaru and Gija people, until a 1983 documentary introduced this UNESCO-listed wonder to the world. But wandering among these 350-million-year-old sandstone formations is far from the only highlight of visiting Purnululu National Park, with walks taking you to the likes of Echidna Chasm, which lights up in brilliant shades of tangerine and scarlet when the sun is overhead, and the natural amphitheatre of Cathedral Gorge.

Reached via a 4WD-only access road off the Great Northern Highway, the park has two campgrounds. Towards the northern end (for attractions including Echidna, the Bloodwoods and Stonehenge), Kurrajong Campground has 100 grassy campsites with some shade. Further south, closer to Cathedral Gorge and the domes, the Walardi Campground offers 37 sites in similar terrain – you'll pass the Purnululu Visitor Centre on your way to both. South of Walardi, the Bungle Bungle Savannah Lodge offers ensuite cabin accommodation with meals, while the Bungle Bungle Wilderness Lodge is used by APT tours. Flight-seeing tours landing nearby add some noise, but when the last one leaves for the day, a deep sense of calm descends.

🔵 THE PITCH

Camp within a short drive of the mesmerising sandstone domes of the Bungle Bungle Range, with two bush campgrounds in the national park to choose from.

When: May–Oct
Amenities: untreated water, toilets, picnic tables
Best accessed: by car (4WD)
Cost: $13
Contact: https://exploreparks.dbca.wa.gov.au

EL QUESTRO
KIMBERLEY

09

The name given to this vast Kimberley property by its first non-Indigenous owner conjures up images of an exotic desert ranch, and that's not far from what you get at El Questro, a sprawling wilderness of gorges and grasslands just off the northern end of the famous Gibb River Road.

The cattle station on the lands of the Wanjina Wungurr Wilinggin and Balanggarra peoples was transformed into a tourism destination in the 1990s, with three accommodation centres: Emma Gorge with glamping tents in easy reach of the beautiful Emma Gorge waterfall; the Station with rooms, permanent tents and camping; and the Homestead with luxurious clifftop suites.

Nestling alongside the Pentecost River, the shady campsites in its main campground are quite close together (book well head for the handful of private sites), yet offer some of the best facilities in the Kimberley, complete with a croc-free waterhole perfect for relaxing in after bumping along the Gibb. Grab a map from reception and spend your stay exploring the 2800 sq km (1081 sq mile) property's ancient gorges and lush springs on self-guided hikes, or sign up for a range of guided tours, from horse trekking to cruises. You'll need a 4WD to access El Questro.

 **THE PITCH**

This former cattle station offers an incredible base for exploring the dramatic natural features of the Kimberley, with camping and glamping options to suit all budgets.

When: May–Oct
Amenities: drinking water, toilets, showers, wi-fi, camp kitchen, laundry facilities, electricity, cafe, store, fuel
Best accessed: by car (4WD)
Cost: camping from $59, plus $23 visitor permit
Contact details: www.elquestro.com.au

CAMPING WITH CUSTODIANS
VARIOUS LOCATIONS

There are more than 200 remote Aboriginal communities dotted across Western Australia. In an Australia-first initiative, the Western Australian Government has so far partnered with half a dozen of them to develop high-quality campgrounds as part of its Camping with Custodians programme. Offering transformative opportunities for cultural exchange, the campgrounds are located in some of the most scenic and interesting corners of the WA outback.

If you're planning a road trip on the Kimberley's Gibb River Road, don't miss the Imintji Campground at the foot of the spectacular Wunaamin Miliwundi Ranges, where you can purchase Aboriginal art direct from artists at the community Art Centre. Headed along the Great Northern Highway? Pull into the Mimbi Caves Campground, an hour's drive east of Fitzroy Crossing, which doubles as the launchpad for Aboriginal-led tours of the incredible Mimbi Caves on Gooniyandi Country. Step back into Creation times as your guide shares the evocative Gooniyandi Dreaming story of the blue-tongue lizard and the mud lark over billy tea and fresh damper before heading back to your campsite with a deeper understanding of the world's oldest living cultures.

THE PITCH

Support the Traditional Custodians of the lands you travel across and learn more about their rich cultures by seeking out this excellent collection of Aboriginal-owned campgrounds in remote areas.

When: May–Oct
Amenities: limited untreated water, toilets, camp kitchen, picnic areas, firepits; some campgrounds have showers
Best accessed: by car
Cost: from $20
Contact details: https://mimbicaves.com.au; https://imintji.com.au

10

© PHILIP SCHUBERT | SHUTTERSTOCK

Wildflowers

Western Australia lays claim to the largest collection of wildflowers on the planet, with 60 per cent of species found nowhere else on Earth. Here's how to take in the botanical spectacle.

More than 12,500 flowering plants burst into bloom across the southern half of Australia's largest state each season, including rare and unique species that have evolved in isolation to survive in nutrient-poor soils. From blush-cheeked donkey orchids to unusual circular wreath flowers, hot-pink hakeas to colourful carpets of everlastings, the rich diversity of flowering species found across Western Australia is a feast for the senses.

Wildflower walks, driving trails, guided tours and festivals offer excellent opportunities to identify and admire the startling beauty of Western Australia's wildflowers, which spring up everywhere from inner-city gardens to national parks, from the sparkling coast to the rugged outback. Visitors can help to safeguard the state's floral splendour by always staying on marked trails, and never picking wildflowers – all of Western Australia's native flora is protected.

WHEN TO GO

Wildflower season typically lasts for six months, kicking off in the northern Pilbara region in June and moving across the Goldfields and down along the Coral Coast. By September, Perth's urban parks – including Kings Park/Kaarta Koomba, home to more than 3000 wildflower species – and the Swan Valley are awash with colour. The season comes to a dazzling close in November in the South West, a biodiversity hotspot also known for the premium wines of the Margaret River

region. Check with local visitor centres for the latest information on where and when wildflowers are blooming, which can vary from season to season.

WILDFLOWER DRIVES

Among the best ways to enjoy Western Australia's flora is to hit the road. Weaving through 'wildflower country' (https:// wildflowercountry.com.au) north of Perth, the 309km (192-mile) Wildflower Way is a great place to start. Tourism Western Australia and the Wildflower Society of Western Australia, as well as regional tourism organisations including Australia's Golden Outback and Australia's South West, have curated a number of memorable wildflower-based driving itineraries – check websites for details.

Wildflowers in Kalbarri National Park (top); Lesueur National Park (above) and its blooming magenta starflowers (left) and banksia (inset)

FIVE TO TRY

Perth parks
Each September the Kings Park Festival hosts a wildflower party within its Botanic Garden.

Leeuwin-Naturaliste National Park
The Cape to Cape Track is lined with wildflowers from September to November.

Lesueur National Park
Look out for blue tinsel lily and magenta starflowers among the 900 species in this park north of Perth.

Kalbarri National Park
Pink pokers, kangaroo paw and other blooms brighten the rugged landscape north of Geraldton from July.

Gascoyne-Murchison region
From July to October this corner of the Golden Outback erupts with colour.

MITCHELL FALLS CAMPGROUND
MITCHELL RIVER NATIONAL PARK

Cascading down a sandstone gorge into a series of emerald pools, Mitchell Falls – known as Punamii-Uunpuu to its Wunambal Traditional Custodians – is now popularly viewed on flight-seeing tours, but there's nothing quite like hiking into this lost-world landscape on your own two feet from the park's only campground.

Getting here is all part of the fun, with the only road access via the 4WD-only Kalumburu Road and Mitchell Plateau Track; a total distance of 539km (335 miles) from Kununurra or 869km (540 miles) from Broome. Then it's an 8.6km (5.3-mile) return walk to the falls. The trail from the campground passes Little Mertens Falls, where you can also swim (allow 30 minutes return). The shorter River View Walk, meanwhile, takes you to a lookout with views over the Mitchell River.

With a helipad opposite the basic, non-bookable campground (bring cash), which despite its remoteness can get busy, camping here can be a little noisy. But it's worth it to experience this surprisingly lush corner of the Kimberley, where fan palms rise up alongside eucalypt forests.

While there is no entry fee to this national park, you will need an Uunguu Visitor Pass to visit Mitchell Falls. Firewood collecting areas are signposted along Port Warrender Road.

11

💬 THE PITCH
Visit the Kimberley's most famous waterfall on a gorgeous bushwalk from its namesake campground, which can be combined with a heli ride for a birds-eye view.

When: May–Oct
Amenities: toilets, some sites with fire ring
Best accessed: by car (4WD)
Cost: $11
Contact: https://explore parks.dbca.wa.gov.au; https://wunambal gaambera.org.au

BIG LAGOON CAMPGROUND
FRANCOIS PERON NATIONAL PARK, SHARK BAY

12

Shark Bay scored its World Heritage status for its rich and vast seagrass beds, its dugong population, and its stromatolites (ancient algae colonies which form distinctive hard, dome-shaped deposits). But it's also a cultural landscape, with about 130 registered Aboriginal sites, as well as relics from its pastoral era. At its heart, some 850km (528 miles) north of Perth, is a historic sheep station-turned national park occupying the top half of the Peron Peninsula. Shark Bay's scarlet sands and aquamarine waters meet with dramatic effect at its Big Lagoon Campground, the most scenic of Francois Peron National Park's five camping areas with well-spaced campsites and a sheltered deck near the shoreline.

The bay's calm waters are perfect for kayaking and stand-up paddleboarding – pair wildlife-spotting with Aboriginal culture on a paddling tour with Nhanda and Malgana man Darren 'Capes' Capewell of Wula Gura Nyinda Eco Cultural Adventures. You may see dugongs, dolphins, turtles and even some of the UNESCO site's namesake sharks – but don't fret, it's safe to swim here.

With the sealed road ending at the Peron Heritage Precinct near the park entrance, you'll need a 4WD to reach all campgrounds. Across the water from Big Lagoon you'll see Dirk Hartog Island, a similarly excellent camping location accessed via ferry and 4WD.

🔵 THE PITCH

Camp amid some of nature's most eye-popping landscapes in Australia's first World Heritage Site – a unique aquatic playground with a rich cultural heritage.

When: Apr–Oct
Amenities: toilets, barbecues
Best accessed: by car (4WD)
Cost: $11
Contact: https://exploreparks.dbca.wa.gov.au

FITZGERALD RIVER NATIONAL PARK
GOLDEN OUTBACK

13

A string of sublime turquoise bays connects the ruggedly beautiful coastline of Western Australia's Golden Outback in 'the Fitz'. One of Australia's most botanically significant national parks, it's home to nearly 1900 or approximately 15 per cent of the state's described plant species, 75 of which are found nowhere else. Add to that 22 mammal species, 41 reptile species and more than 200 bird species, and you can bank on plenty of wildlife action. And the national park's pair of 2WD-accessible campgrounds put you right in it.

Tucked in the coastal scrub behind its namesake beach at the park's eastern end, Four Mile Campground has 15 privately situated sites, five for tents only. Facilities are top-notch, and from here it's only a short drive to the East Mt Barren trailhead, a handful of beautiful bays (tiny West Beach is particularly pretty) and Cave Point – the start point for the gorgeous coastal Hakea Trail (23km/14.3 miles one-way).

The older-style St Mary Inlet Campground in the park's west offers direct access to the lovely Point Charles Bay, where southern right whales can be seen sheltering with their calves during winter. From the campground it's only 2km (1.2 miles) to the lookout at Point Ann for an elevated view.

🔵 THE PITCH

Walks, wildlife, beaches and blooms await in this epic coastal wilderness between Albany and Esperance, with two national park campgrounds offering a convenient base for exploration.

When: year-round
Amenities: toilets and picnic tables, plus showers, barbecues and shelters at Four Mile Campground
Best accessed: by car
Cost: $11
Contact: https://exploreparks.dbca.wa.gov.au

HIDDEN CABINS
VARIOUS LOCATIONS NEAR PERTH

Even Western Australia's most remote campgrounds can get crowded, especially on weekends and during school holidays. When you'd rather not cross paths with another soul – except perhaps a travelling companion – Hidden Cabins has you covered. A relatively new member of the state's ever-expanding tiny accommodation family, its three off-grid cabins are configured to complement their unique and secluded locations on private properties within two hours of Perth.

Florence nestles in the peaceful eucalypt forests of the Peel region, while Henry is flanked by majestic jarrah, marri and yarri trees and overlooks the rolling vineyards of the Ferguson Valley. The latest addition, Margot, sits on a dairy-turned-regenerative farm, also in the Ferguson Valley. Unlike the other two one-bed cabins, Margot sleeps four, allows dogs, and has an outdoor bath. Perfect for a winter escape, all cabins have a small wood-burning stove, a firepit outside, and a curated selection of books and board games to entice you from checking your phone. Huge windows that invite nature in are another common feature. Adding zing to the adventure is the secrecy of the locations, revealed just three days before your stay.

🔌 THE PITCH

When you need to unplug beyond the madding crowds, this trio of tiny cabins offer privacy and comfort in re-energising and easy-to-reach nature-based settings.

When: year-round
Amenities: drinking water, private bathroom, kitchenette, electricity
Best accessed: by car
Cost: from $300
Contact details: www.hiddencabins.com.au

14

© ANNA REES PHOTOGRAPHY

KIMBERLEY DREAMING
KUNUNURRA

Twisting and turning through dramatic red gorges before discharging into the Cambridge Gulf, the mighty Ord River is one of the Kimberley's most scenic waterways. And it sets the scene for an intimate getaway on the *Kimberley Dreaming*, a two-bedroom houseboat licenced to sleep six. Guests can choose from three scenic mooring locations, each more adventurous than the next.

With easy access to Kununurra by kayak, Lily Creek Lagoon offers superb birdwatching. Some 20km (12 miles) upriver, Wallaby Cliffs puts you on the edge of the Carr-Boyd Range, where wallabies graze on cliffs and archerfish dive for their dinner as you relax on board. Another 10km (6 miles) upriver, the Sandy Creek location immerses you in the wilderness of the Ord's upper reaches, where the night sky shines even brighter, and the howls of dingoes echo through the gorges. While you're not allowed to move the boat, you can paddle kayaks along the river to look for crocodiles basking in the sun, and even sleep al fresco under a mozzie net on the deck. Simply BYO food, drinks and a fishing rod, and farewell the mainland for at least three days of off-grid fun on the water.

🌀 THE PITCH

Cast away from Kununurra for one of Western Australia's most unique off-grid overnight experiences aboard a houseboat moored on a peaceful stretch of the Ord River.

When: May–Oct
Amenities: drinking water, private bathroom, kitchenette, electricity
Best accessed: by boat
Cost: $2200 for 3 nights
Contact: www.kimberleyhouseboats.com

15

© CRISCOGRAPHY | SHUTTERSTOCK

16

WOLFE CREEK CRATER CAMPGROUND
WOLFE CREEK METEORITE CRATER NATIONAL PARK

The inspiration for the terrifying Australian horror movie *Wolf Creek*, the real Wolfe Creek Crater has an eeriness to it without its association with a fictional serial killer. Formed by a 50,000-tonne meteorite smashing into the desert around 300,000 years ago, the immense crater, measuring 880m (2887ft) across, wasn't discovered by Europeans until 1947. But it has been known to its Jaru Traditional Custodians and other Aboriginal people for much longer, featuring in several Creation stories.

There are ten non-bookable sites (bring cash) off the gravel road into the park, some with minimal shade. It's just over 600m (1968ft) from the campground (or 200m/656ft from the crater car park) to the crater rim – a steep, rocky climb, ideally timed for sunrise or sunset, when the rim glows a deep marmalade hue before darkness creeps in. Climbing down inside the crater isn't permitted, but from the rim you'll see the salt pan at its centre. Listen for the cackle of pastel pink Major Mitchell's cockatoos attracted to the paperbark and wattle trees, and keep an eye out for small ringtail dragons basking on the rocks.

💬 THE PITCH

Take a long, dusty detour off the Great Northern Highway to camp beside the world's second-largest meteorite crater. Yes, the one from *that* movie.

When: May–Oct
Amenities: toilets
Best accessed: by car (ideally 4WD)
Cost: $8
Contact: https://exploreparks.dbca.wa.gov.au

BIBBULMUN TRACK HUTS
SOUTHEAST OF PERTH

Yellow markers symbolising the Waugal, the Rainbow Serpent in Noongar Aboriginal Dreaming, show the way on the epic, 1000km (621-mile) Bibbulmun Track. From Kalamunda on the outskirts of Perth to the south-coast town of Albany, this is the state's longest walking trail, with all 49 camping areas reachable only on foot.

Camping is only allowed at the designated sites dotting the trail, which has nine sections and takes between six and eight weeks to complete. Each camp has a three-sided hut (except for the fully enclosed hut at Mt Wells) accommodating up to two dozen walkers, but as they can't be booked, all overnight hikers need to carry a tent. The original huts are wooden, but more recently rebuilt huts, including the popular Helena Campsite in the Perth Hills (section one), are made from rammed earth, known for its fireproofing and insulation properties. Other crowd favourites include Waalegh Campsite (section one) for its views, Dog Pool Campsite (section seven) for its freshwater rock pools and Boat Harbour Campsite (section eight) for its easy access to lovely Boat Harbour Beach.

It's free to hike the track, but buying a guide or map from the Bibbulmun Track Foundation is recommended.

🧭 THE PITCH

With bushfire-affected walkers' huts on this epic long-distance trail now being rebuilt with hardier materials, camping on the Bibbulmun Track is becoming (slightly) more comfortable.

When: Apr–Nov
Amenities: toilets, untreated water
Best accessed: by bus or trailhead transfer
Cost: free
Contact details: www. bibbulmuntrack.org.au

17

FRANKLAND RIVER

© STEVE WATERS

STARGAZERS CAMPGROUND
MILLSTREAM CHICHESTER NATIONAL PARK

The name of this medium-sized campground nestling in spinifex and eucalyptus woodland reveals what you can look forward to when darkness falls in Millstream Chichester National Park, a rugged landscape of rocky escarpments and tree-lined watercourses some 120km (75 miles) south of Karratha. With 15 unpowered gravel-surface campsites of varying sizes in six areas, Stargazers Campground is also the closest place to camp to the park's top attraction: Nhanggangunha (Deep Reach Pool).

You can swim and launch canoes, kayaks or stand-up paddleboards in this serene natural spring-fed pool, but the local Yindjibarndi people ask that you treat this sacred place, home to the Warlu, or Rainbow Serpent, with respect. Learn more about the park's cultural significance and pastoral history at the Millstream Homestead Visitor Centre, less than 3km (1.9 miles) north of the campground.

June and August are great months to visit, when the red plains erupt with colourful wildflowers; look out for vivid red Sturt desert peas, cheerful northern bluebells, and even several species of hibiscus in this remote desert oasis. The park's permanent source of water also attracts myriad birds such as the rainbow bee-eater, sacred kingfisher and blue-winged kookaburra, as well as 12 species of raptor. Wheelchair-accessible toilets are available at Nhanggangunha and the Miliyanha Campground, near the homestead.

18

🗨 THE PITCH

Paddle in a sacred natural desert pool by day and stretch out under a blanket of stars by night in this former pastoral station–turned national park.

When: Apr–Oct
Amenities: toilets, limited untreated water, barbecues, picnic tables
Best accessed: by car
Cost: $11
Contact details: https://exploreparks.dbca.wa.gov.au

NEW ZEALAND NORTH ISLAND

Embrace the sights, sounds and smells of the North Island's beautiful wilderness, from volcanic mountains to sandy beaches in the winterless north.

Best time: Sep-May (camping/glamping), Jun-Aug (cabins)
Best national parks for camping: Tongariro National Park, Te Urewera, Coromandel Forest Park
Best camping trails: Te Araroa Trail, Waiohine Gorge, Kauaeranga Kauri Trail, Pureora Timber Trail
National parks pass required: No but see the South Island, page 220
Useful contacts: Department of Conservation (DOC) www.doc.govt.nz; www.tiakinewzealand.com

New Zealand is a major international draw for lovers of the great outdoors. Trampers (hikers) head up mountains and along rivers and coasts with backpacks, tents and hiking boots to experience the country's unique landscapes. Others explore with two wheels on long-distance rides like the Pureora Timber Trail and the North Island's Coast to Coast cycle trail. There are also plenty of places to get out on the water here, including Lake Tarawera Hot Water Beach where you arrive for your overnight camp by boat, and the campsites on Urupukapuka in the Bay of Islands – accessed by ferry and foot. Even one of New Zealand's Great Walks involves a multi-day canoe journey along the Whanganui River.

Don't want to carry all your gear to summit mountains or ford streams? Thankfully New Zealand's North Island has plenty of wild places to camp that can be accessed by car (or campervan). These simple campsites, usually managed by the Department of Conservation (DOC), may be inexpensive but are in priceless locations. Beyond the campsites, the North Island has an increasing number of self-contained cabins overlooking farmland and vineyards. From the old-school beach 'bach' to newly renovated cabins with luxury touches like outdoor bathtubs, gourmet kitchens and organic products, you can enjoy the North Island without compromising on comfort.

FREE CAMPING

There are limited locations for 'freedom camping'. Most DOC (and all privately owned) campsites are bookable with a small charge per night.

SUPPLIES

Big outdoors brands like Kathmandu and Macpac are located in Auckland and some smaller cities. Plus, there are army surplus stores and camping sections at Kmart and The Warehouse.

SAFETY

Seek treatment if you get bitten by the (endangered) katipō spider. Stay abreast of weather, earthquake and volcano conditions. Don't walk off marked tracks.

BEST REGIONS

Bay of Islands & Northland
Camp by pristine beaches in easy reach of the gourmet regions. Wander pockets of ancient kauri forests and experience New Zealand's enduring Māori culture.

Twilight at Rua Awa Lodge (left); the suspension bridges of the Timber Trail

Taupō & the Central Plateau

Beyond the imposing (and occasionally erupting) volcanic mountains Ruapehu, Tongariro and Ngauruhoe on the North Island's central plateau, experience New Zealand's longest river, the Waikato and the adventure capital of Lake Taupō.

Waikato & the Coromandel Peninsula

Dig yourself a natural sandy spa pool, hike to the top of the Pinnacles, and splash in the sparkling Wentworth River. The Coromandel has popular beaches, regenerating forests and historic gold-mining towns to explore.

© JOEL MCDOWELL; © EPIC CYCLYE ADVENTURES

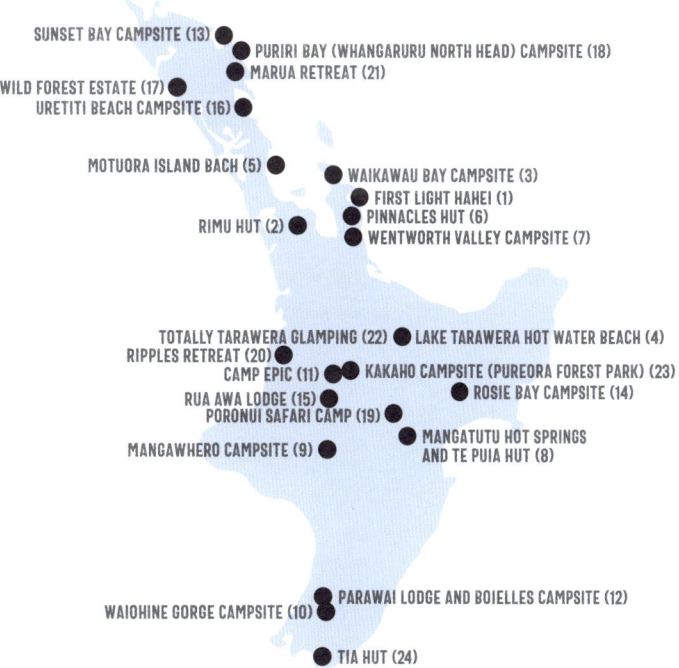

SUNSET BAY CAMPSITE (13)
PURIRI BAY (WHANGARURU NORTH HEAD) CAMPSITE (18)
MARUA RETREAT (21)
WILD FOREST ESTATE (17)
URETITI BEACH CAMPSITE (16)
MOTUORA ISLAND BACH (5)
WAIKAWAU BAY CAMPSITE (3)
FIRST LIGHT HAHEI (1)
PINNACLES HUT (6)
RIMU HUT (2)
WENTWORTH VALLEY CAMPSITE (7)
TOTALLY TARAWERA GLAMPING (22)
LAKE TARAWERA HOT WATER BEACH (4)
RIPPLES RETREAT (20)
CAMP EPIC (11)
KAKAHO CAMPSITE (PUREORA FOREST PARK) (23)
RUA AWA LODGE (15)
ROSIE BAY CAMPSITE (14)
PORONUI SAFARI CAMP (19)
MANGATUTU HOT SPRINGS
AND TE PUIA HUT (8)
MANGAWHERO CAMPSITE (9)
PARAWAI LODGE AND BOIELLES CAMPSITE (12)
WAIOHINE GORGE CAMPSITE (10)
TIA HUT (24)

FIRST LIGHT HAHEI
COROMANDEL PENINSULA

Perched in the foothills beyond a popular campground at Hahei on the Coromandel Peninsula, you'll find the newly built canvas glamping accommodation at First Light Hahei. Enjoy sunrise over breakfast with views of the islands dotted through Mercury Bay, or take in the moonrise from your outdoor bathtub. Hahei Beach is famed for its spectacular pōhutukawa trees and pink sand, with families returning year after year to fish, beachcomb, and swim. It's also the gateway to the Insta-famous Cathedral Cove.

First Light is fully off-grid, running on solar power only. There is mobile reception and wi-fi, with a battery bank to recharge devices. The outdoor terrace kitchen has a small fridge plus all the equipment you need to fully self-cater, plus a supply of plunger coffee, tea and milk. In the ensuite bathroom, there's complimentary toiletries, plus fluffy towels. Check in at the resort, then it's five minutes on a 4WD track up to the site. Make sure you pack for cool nights even if you're here in the height of summer. At the larger campground resort (a 20-minute walk away) you'll find barista coffee, a seasonal bar, and lunch and dinner options. Hahei village also has a general store, restaurants and a brewery.

 **THE PITCH**

This new private luxury glampsite is set on a hill overlooking the ocean, and managed by Hahei Beach Resort.

When: year-round
Amenities: outdoor terrace kitchen, gas cooktop, BBQ, outdoor bath, private bathroom, wi-fi
Best accessed: by 4WD (Hahei Beach Resort can also arrange access)
Cost: from $249
Contact: www. haheiholidays.co.nz/ first-light-hahei

01

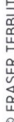
© FRASER TEBBUTT

RIMU HUT
ARARIMU, NEAR AUCKLAND

Originally hand-built for the owner's grandchildren to indulge in sleepovers and nature-based adventures, Rimu Hut is styled on a tramper's hut, with leafy views from the forest bathroom and bed spaces alike. This magical spot is evocative of secret childhood cubbies: tucked away from the modern world, it smells of hewn timber and the petrichor of rain-soaked forest. Essentially one room with a loft space, it comfortably sleeps five, with three single mattresses upstairs and two built-in beds downstairs. There's a nearby compost toilet and adjacent bathroom, plus a gas hob for cooking and a small

wood-fired stove to keep toasty warm in winter.

This is a truly rustic experience – the hut has no power, just torchlight or candlelight after dark, a food safe and, of course, no wi-fi (or television). Entertain yourself

(and your family) with games and various walking trails. A wetland walk reveals a delightful pond and the fairy-lights of the region's night-time glow worms.

This is a popular spot for a retreat close to Auckland. A hand-crafted treehouse on the property can be made available if Rimu Hut is booked out; ask the owners for more info.

  **THE PITCH**

Venture just beyond South Auckland for a rustic retreat in an off-grid A-frame chalet on the edge of native forest.

When: year-round
Amenities: compost toilet, gas hob, food safe (no fridge)
Best accessed: by car
Cost: $200
Contact: www.airbnb.co.nz

03

WAIKAWAU BAY CAMPSITE
COROMANDEL PENINSULA

On the northern tip of the Coromandel Peninsula, you'll find this popular off-the-beaten-track campsite overlooking the stunning white-sand surf beach of Waikawau Bay. With two estuaries and a significant wetland, the area is a haven for native birds. It's also within walking distance of sheltered Little Bay, which is best for more sedate water activities like stand-up paddleboarding – and kids can splash around here away from rolling waves. Mobile phone coverage is poor, but a public payphone winds back the clock – for anyone born this century, placing a call may feel like a puzzle to solve.

Though popular, paddock after paddock allows relatively private camping (except in the peak of summer). The showers are cold; you need to bring your own cooking equipment; and the nearest grocery store is the super-quaint Colville General Store (a winding 25-minute car ride away). Stop in Coromandel town (an hour away) to stock up your camp larder – Little Bay has a small residential community, but you'll need to be fairly self-sufficient. In summer, a small shop opens at Waikawau in an old-school house, selling milk, ice, bread, and tantalisingly cold ice creams.

 THE PITCH

A large campsite with direct access to a pristine surf beach and an isolated, back-to-nature feel – albeit with cold ice-creams on demand in summer.

When: year-round
Amenities: powered and unpowered sites, cold showers, pit toilet
Best accessed: by car
Cost: $7.50–$21
Contact: www.doc.govt.nz

© THOMAS HAGENAU | SHUTTERSTOCK

LAKE TARAWERA HOT WATER BEACH
ROTORUA AND BAY OF PLENTY REGION

Tarawera Hot Water Beach is on a walking trail that meanders through stunning native forest by spectacular jade waters. But better than tramping here, follow in the footsteps of the 19th-century European tourists and cruise across the lake. They came to marvel at one the world's unofficial wonders: the Pink and White Terraces, a series of spectacular thermal pools descending in a cascade of rosy-hued silica steps. Sadly, in 1886, Aotearoa's deadliest volcanic eruption destroyed this incredible natural feature, though the area remains popular for its natural hot springs.

At this simple campground, you'll find 30 unpowered tent sites in a small clearing by one of the lake's hot water beaches. There's a pit toilet and cooking shelter (bring your own camp stove, no fires permitted). Bookings are essential, and if you bring your own boat or canoe, be aware the mooring spots are limited. Another few kilometres' walk takes you to a lookout over neighbouring Lake Rotomahana ('warm lake' in te reo Māori). These are sacred places for which the Tiaki Promise – a commitment to care for New Zealand, now and for future generations – is essential. That means following instructions like cleaning your gear between lakes to avoid the spread of freshwater pests.

04

💬 THE PITCH

Set up camp and hand-dig a natural spa bath at low tide at Te Rātā Bay, before exploring the walking tracks around sacred Lake Tarawera.

When: Sep-May
Amenities: unpowered tent sites, cooking shelter, pit toilets, untreated water
Best accessed: by water taxi, canoe, walking
Cost: $7.50–$15
Contact: www. totallytarawera.com

MOTUORA ISLAND BACH
AUCKLAND REGION

Accessible only by boat (or sea kayak if you're game), Motuora Island is a large island reserve in the western Hauraki Gulf. Historically used as farmland, the extensive planting of native vegetation and the eradication of non-native pests will hopefully restore its original forest cover over time. For this reason, walkers are asked to stick to the paths, to avoid walking on newly planted vegetation or near bird nesting areas. Motuora Island is also one of New Zealand's 'kiwi creche' locations, part of Operation Nest Egg, which raises kiwi chicks until they are big enough to deal with non-native stoats, before releasing them on the mainland.

At the back of the 36-person campsite is a bookable beach bach (that's Kiwi for a basic holiday house), sleeping up to four people. From here, campers (and bach-dwellers) make the most of their very own harbour island. Look out for the New Zealand dotterel, shore plovers and majestic grey-faced petrels on birdwatching walks, or swim, snorkel and kayak from the gently sloping beach.

Water taxis are available to access the island. You'll need to bring all food and drink and take everything with you on departure. Don't forget insect repellent.

🪙 THE PITCH

Snorkel by day and listen for brown kiwis in the undergrowth at twilight, before drifting off to sleep at an old-school beach bach on this island sanctuary.

When: year-round
Amenities: basic kitchen, solar power, hot showers
Best accessed: by boat
Cost: cabin $105
Contact: www.doc.govt.nz

© VEE SNIJDERS | SHUTTERSTOCK

PINNACLES HUT
COROMANDEL PENINSULA

06

The historic Kauaeranga Kauri Trail (6km/3.7 miles one way) up to the serrated outcrop of the Pinnacles can be tackled in a long day-walk (returning via the Billygoat Track), but it's even better with an overnight stay at Pinnacles Hut. From the hut you can return to those dramatic boulders to take in the 360-degree views of a vivid sunset or (sometimes misty) sunrise – without the crowd.

The trail up to Pinnacles Hut is not an easy stroll. Trampers cross the Kauaeranga River on a swing bridge before trudging up a fairly steep ascent, following in the footsteps of the pack horses that once served kauri loggers. Rain and the risk of slipping are a bit of a constant here, even if the region is otherwise usually mild to warm. Campers need to bring their own provisions as well as water, sleeping gear and cookware. However, the facilities at Pinnacles Hut are beyond the average DOC hiking shelter, with gas cookers, solar-powered lighting and a solid-fuel burner for cold nights. Yes, it's big, with 80 bunks (pack earplugs if you're a light sleeper), but the onsite wardens contribute to a warm convivial vibe.

🔵 THE PITCH

Beyond Coromandel's famous beaches, on the Kauaeranga Kauri Trail in the Coromandel Forest Park, you'll find a well-run backcountry hut with views of the Pinnacles.

When: year-round
Amenities: drop toilet, cooking facilities, rainwater tanks, heating in winter
Best accessed: by foot
Cost: $12.50–$30
Contact: www.doc.govt.nz

© BETANEDVED | SHUTTERSTOCK

WENTWORTH VALLEY CAMPSITE

COROMANDEL FOREST PARK

On the site of a former gold mining settlement, which included houses, shops and schools, you will find this secluded campsite owned by the DOC but run by a private operator. The site has a mix of private spots snuggled among ferns, and a flat open area which is often favoured by families where kids can run free. The Wentworth Falls Walk is an easy hike (around 3km/1.8 miles return) featuring two bridges to cross and well-kept tracks through North Island silver ferns. More experienced hikers can climb the steeper tramping track to the icy basin at the top of the falls. Traverse the range from Wires or Maratoto Track and spy views of the Coromandel Peninsula below. Off the local walking tracks are hobbit hole-like abandoned mining shafts (so don't let your kids – or dog – wander into the unknown). For something less vigorous, there are also a number of swimming holes along the Wentworth River for a relaxing dip in the crystal-clear water.

Powered sites need to be booked in advance year-round, but for unpowered camping you can just turn up (except during school holidays). There are no bins – take all your rubbish out with you. And open fires aren't allowed, so BYO camp stove or gas cooker.

THE PITCH

A well-maintained campsite in native bush alongside Wentworth River, with easy access to swimming holes and the spectacular Wentworth Falls.

When: Dec–Apr (book ahead), May–Nov (bookings for powered sites only)
Amenities: gas BBQs, fresh water, non-flush toilets, hot showers ($2 coin), some powered sites
Best accessed: by car (beware: roads can flood)
Cost: $15–$21
Contact: www.went worthvalleycamp.co.nz

07

08

MANGATUTU HOT SPRINGS AND TE PUIA HUT
HAWKE'S BAY

Kaweka Forest Park has a number of treks, from short day walks to multi-day loops among the mixed forest of beech trees and historic pine plantations. Between hikes, you can cool off with a dip in the Mohaka River (also used for Level 2 to Level 5 white-water kayaking or rafting) or try your hand at fishing for trout. However, the major draw to the region for campers is the free thermal hot springs to soak in after a long day. At Mangatutu Hot Springs, two pools are built on a deck overlooking the steep river valley and forested hills. Thermal water trickles into the pools from a deep hole near the top terrace.

Camping is at the end of a gravel road on a first-come, first-served basis. Not too far away (8km/5 miles), Te Puia Hut is an excellent overnight stay for families with children or anyone new to hiking. There you'll find a 26-bed bunkhouse with a wood burner for winter. If it's full, set up camp outside. It's another 30-minute walk along the river valley to a second set of free hot springs.

🏕 THE PITCH

Very popular campground above the Mohaka River in the Kaweka Forest Park, near free hot springs with stunning views.

When: Oct–Apr
Amenities: unpowered tent sites, untreated water, pit toilets, a shelter
Best accessed: by car
Cost: free
Contact: www.doc.govt.nz

MANGAWHERO CAMPSITE
TONGARIRO NATIONAL PARK, CENTRAL NORTH ISLAND REGION

If the term 'New Zealand' conjures dramatic landscapes and peculiar geological phenomena, Tongariro National Park is that New Zealand. The large park is home to three volcanoes (Tongariro, Ngauruhoe and Ruapehu), emerald lakes, ancient lava flows, steaming craters, colourful silica terraces and soft alpine meadows. *Lord of the Rings* fans can literally reach Middle Earth from here: a 10km (6-mile) walk from the campsite takes you to Mangawhero Falls, where you can imagine Gollum trying to catch fish in the scene from *The Two Towers*. In winter, icicles form spectacularly around the cascade.

The campsite comprises 12 unpowered tent sites and six for campervans, with drop toilets and picnic tables. You'll have to BYO camp stove for cooking and boiling water as fires aren't allowed, but it's not too far from Ohakune township, for a delightful walk into town for coffee, food and shopping. As well as the waterfalls, there's a kid-friendly forest hike nearby, and plenty of spots for a refreshing paddle in the river. You can also hike up to Mangaehuehu Hut (bookings required to overnight there) which is set among the tussock grasses with views of Mt Ruapehu, an active volcano and the North Island's highest peak.

💬 THE PITCH

Camp at the base of snow-capped Mt Ruapehu — with forest walks, swimming spots, and a *Lord of the Rings* filming site all within reach.

When: Oct-Apr
Amenities: drop toilets, untreated water, picnic tables
Best accessed: by car
Cost: $7.50-10
Contact: www.doc.govt.nz

09

© MARTIN VLNAS | SHUTTERSTOCK

© ALAN BENGE | SHUTTERSTOCK

WAIOHINE GORGE CAMPSITE

TARARUA FOREST PARK, WAIRARAPA

At the head of two gorges, this basic campsite is positioned in a beautiful bush setting near a river junction. From here, you cross a breathtaking swing bridge to the Holdsworth–Kaitoke walking track. Access to the campsite is via a gravel road which is quite tight – a potential challenge for campervans. Check road conditions before departing, as heavy rains can make the final section accessible only on foot.

Nearby tramping trails (of varying difficulty) provide a chance to explore the area's wild beauty up close. An excellent circular day walk takes you to the historic Cone Hut, where

it is also possible to overnight (for free, no booking required). There's also an abseiling site nearby, but you need to contact the DOC to arrange access. For most visitors, however, the main focus here is the Waiohine River.

As well as swimming, fishing, canoeing and kayaking, campers can wade through the river to a secret beach and float back down to the bridge. Deeper rock pool areas entice brave souls to leap into the clear waters from the rocks – always check for obstructions before jumping into a river. Hit Greytown for supplies (including insect repellent).

  THE PITCH

Wade upstream to a secret beach from this bush campsite in the Tararua Forest Park, before floating back down to the nearby swing bridge.

When: year-round
Amenities: unpowered tent sites, flush toilets, seasonal firepit, untreated water
Best accessed: by car
Cost: $5-$10
Contact: www.doc.govt.nz

CAMP EPIC
PUREORA FOREST PARK

A dream track for the experienced cyclist, the Timber Trail is a two-day 82km (51-mile) mountain-biking adventure through the Pureora Forest, ascending Mt Pureora and crossing four huge suspension bridges. Graded easy to intermediate (by mountain-biking standards), this is not a ride to tackle without proper kit, a decent bike and a base level of cycling fitness. However, for those who put in the effort, the rewards are epic (as is the crew that support you).

Halfway through the trail, you can overnight at the well-presented Camp Epic. The camp is run by Epic Cycle Adventures, who offer a well-priced package that includes transporting luggage to the campsite while you work up an appetite on the trail. Situated in a forest clearing, 16 canvas tents share a firepit, hot showers, and everything you need to cook in an open-air communal kitchen. Campers need to bring food and drink, and should note that phone coverage is sporadic (although there is solar recharging for your electronic devices). While you're supported on this trip, you nonetheless need to be fairly self-sufficient. Book well in advance for October to May. And if you're travelling away from home, you'll be pleased to know e-bike and mountain-bike hire is also available.

 **THE PITCH**

Along a once-remote historic logging tramway, Camp Epic is a glamping respite from two days of adventure cycling along the Timber Trail.

When: year-round
Amenities: showers and toilets, communal kitchen, picnic tables
Best accessed: by bike
Cost: $40 BYO tent; glamping $120
Contact: www.thetimbertrail.nz

© HANS WISMEIJER | SHUTTERSTOCK

PARAWAI LODGE AND BOIELLES CAMPSITE
KAPITI COAST REGION

A short walk from the Boielle Flat carpark, near the confluence of the Waiotauru and Otaki rivers, is the neat 18-bunk Parawai Lodge, the perfect base from which to explore Tararua Forest Park. It's first-in, first-served, so bring your own camping gear in case it's full. Alternatively, set up in one of 15 unpowered site on the Boielles Campsite, right by the river. At the time of research, access to the Ōtaki Forks area was limited due to a landslip, but pedestrians could access an alternative route across private land.

Once here, you have the possibility of river swimming, rafting and canoeing these two pristine rivers, as well as tackling hikes and kid-friendly walks. Taking you through the regenerated forest, the 50-minute Arcus Loop rewards with a wonderful view of the campground and river below. For the more adventurous tramper, the 25km (15.5-mile) Pukeatua Track climbs to a summit through mixed forests. It is also on the Te Araroa Trail – a 3000km (1864-mile) route from the tip of the North Island to the most southerly point of the South Island.

  THE PITCH

Set up camp at this (usually) very accessible spot in the Waiotauru Valley, just over an hour from Wellington, and swim, raft and hike in the Tararua Range.

When: Oct–Apr
Amenities: bunks, unpowered tent sites, untreated water
Best accessed: by car, hiking
Cost: $5–10
Contact: www.doc.govt.nz

SUNSET BAY CAMPSITE
BAY OF ISLANDS

Sunset Bay on Urupukapuka in the Bay of Islands is postcard-worthy. With only two (bookable) tent sites, sleeping up to eight people, you'll fall asleep to nothing but the sound of the wind in the trees and the waves lapping on the shore. Two other sites on the island are also available at Cable Bay and Urupukapuka Bay – each less than a 25-minute walk from Otehei Bay, where the ferry comes in. There's also a cafe at Otehei Bay for proper coffee, but the closest places for provisions are Paihia and Russell, back on the mainland, and you'll need to take all your rubbish to a refuse station in town.

Urupukapuka Island is a treasure to be explored by foot or watercraft. A 7.3km (4.5-mile) walk circumnavigates the island, taking you over forested ridges and past steep cliffs with awesome views (look out for dolphins), then dropping into sheltered bays. Numerous archaeological sites on the island tell some of the pre-colonial story, including Māori Pa sites. Project Island Song is working to restore native habitats (and birdsong). To support their work, bring plant-based detergents only, and follow all biosecurity guidance to keep the island free of pests, including non-native plants or seeds.

🔵 THE PITCH

If privacy is the new luxury, then this two-site camp perched above a secluded beach – on an idyllic semi-tropical island – is as good as it gets.

When: year-round
Amenities: drop toilets, untreated water
Best accessed: by boat and walking
Cost: $9–18
Contact: www.doc.govt.nz

© RAPHI32 | SHUTTERSTOCK

ROSIE BAY CAMPSITE
EAST COAST REGION

The region of Te Urewera is considered a living being under national law, such is the reverence for its lakes and ancient forests. This is Ngāi Tūhoe land – the local *iwi* (tribe), known as 'the Children of the Mist', have a powerful ongoing connection with the region and remain its guardians. The Te Urewera Board grants permits to *manuhiri* (visitors) for activities such as fishing and hunting here.

Small Rosie Bay Campsite here is free, has six unpowered sites and is accessible by road. There's no need to book, but of course you must take care – check conditions before departing, as the campsite is prone to flooding in the off-season.

The campsite also has a boat ramp on Lake Waikaremoana, if you want to bring canoes and head out into the misty lake for a paddle. Not far from the campground is the spectacular Lou's Lookout – a decent ascent past small caves and a 20m (65ft) rock tunnel, but worth it for the view of Panekire Bluffs and the sacred lake from above. It's also close to the start point for one of the North Island's Great Walks, the Lake Waikaremoana Track.

🧭 THE PITCH

Unwind, stargaze, and let the gentle lapping of waves be your soundtrack at this small lakeside campsite at Lake Waikaremoana.

When: Nov–Mar
Amenities: unpowered tent sites, firepit, untreated water, pit toilet, limited mobile coverage
Best accessed: by car
Cost: free
Contact: www.doc.govt.nz/parks-and-recreation

14

© JIRI FOLTYN | SHUTTERSTOCK

Bikepacking

See New Zealand's natural wonders at a sociable and sustainable pace by packing supplies for a few days on a bicycle and pedalling along numerous purpose-built routes.

Don't be put off by the new label for what is an activity as old as the invention of the bicycle itself. Bikepacking, or cycle touring if you prefer, is an unbeatable way of taking a few days (or weeks if you can swing it) to experience the real ups and downs of New Zealand. Over the past decade or two, New Zealanders, led by government policy, have been designing long-distance cycle routes that are intended to provide safe and sustainable avenues into both the nation's most famous regions and also its overlooked corners.

Many of these routes fall under the umbrella of the government-supported Ngā Haerenga New Zealand Cycle Trails scheme, which includes 23 'Great Rides' and a larger number of Heartland and Connector routes. They range from family-friendly daytrips to challenging adventures over hundreds of kilometres. Either way, if you're spending a night away from home you can call yourself a bikepacker, whether you're carrying no more than a credit card for a night in a cabin and cafe meals, or packing a tent, stove and sleeping bag to be fully self-sufficient.

WHERE TO RIDE

New Zealand's North Island generally has a more temperate climate than its South Island. It's also more developed and less mountainous, making it a better bet for novice bikepackers. In the north of the island, explore the Bay of Islands on a bicycle, sandy beaches and the

kauri forests. Around the centre of the North Island, discover a landscape peppered with hot springs and volcanoes. The east coast has several wine regions through which to roam. And around the wet and windy south, there are great rides around Wellington. The complete Tour Aotearoa route, designed by New Zealand's pioneering Kennett brothers, pieces together many of the country's best bikepacking trails and is a fabulous adventure.

Maps of most routes are available from the official Ngā Haerenga New Zealand Cycle Trails website (www.nzcycletrail.com).

Cape Reinga at the very top of the North Island and the starting point of Tour Aotearoa (top); riding kauri country (above); the Timber Trail (left); a break on the Remutaka Loop (inset)

FIVE TO TRY

Mountains to Sea Ngā Ara Tūhono
An official route between Tongariro National Park and Whanganui.

Coromandel Peninsula
Tropical Coromandel is the setting for a weekend away from Auckland with beautiful coastal views.

The Timber Trail
This linear track, with a halfway pitstop, explores regenerating forests in the heart of the North Island.

Remutaka Loop
Escape Wellington on this 145km (90-mile) loop along wild coastline and through Wairarapa wine country.

Tour Aotearoa
A classic north-south route the length of New Zealand. Northern highlights include beaches and kauri forest.

RUA AWA LODGE
CENTRAL NORTH ISLAND

Set on carpet-like lawns, Rua Awa Lodge offers a self-contained cabin for two, and a haven for cooks and gourmands alike. Enjoy a nutritious home-cooked meal ready for your arrival (perfect after a day of hiking), or a grazing platter and bottle of wine on the wood deck. Alternatively, stop by Kakahi general store to collect provisions, get busy in the well-organised kitchen, then dine under the stars in the firepit area.

Birds and wildlife flock here, and there's direct access to Whanganui River to wild swim. If you're into canoeing, rafting or fishing, local outfits can take you out on the water. Mountain bikers are also spoiled for options nearby. For something less adrenaline-pumping, there are yoga and wellness workshops (learn how to make kombucha, nut milk or kimchi, for example) onsite, in the airy yoga studio.

Lodge co-owner Sheryl is a qualified yoga and corrective exercise instructor, as well as a champion for food as medicine and a more environmentally conscious mode of living. Proximity to the Tongariro Alpine Crossing, regarded as New Zealand's best day walk, is another draw for lovers of nature and hiking.

THE PITCH

River access and mountain views are just two of the perks of this well-equipped accommodation with a yoga studio, surrounded by a mix of organic gardens and native bush.

When: year-round
Amenities: TV, DVD, wi-fi, washing machine, bar fridge, gas cooktop, oven, deck, firepit, indoor wood burner, outdoor shower, separate bathroom
Best accessed: by car
Cost: $220
Contact: www.ruaawalodge.co.nz

© JOEL MCDOWELL

URETITI BEACH CAMPSITE
NORTHLAND

Uretiti Beach is part of a 22km (14-mile) expanse of sand along Northland's east coast at Bream Bay – a long shallow indentation, which means the sea is surprisingly warm throughout summer. That could be why this is the location for New Zealand's favourite informal naturist area (head south from the DOC campsite – or north if you want to avoid it). Surfers and body borders will welcome the breaks here; swimmers are advised to swim between the flags. At low tide, shellfish can be yours for the BBQ. And while pets are not allowed at camp, you might see visiting horses from nearby stables galloping along the beach, with islands in the background.

This is a large, accessible DOC site close to Highway 1, but with 300 sites there's plenty of room, and out of the peak season you'll feel like you have the site – and the beach – to yourself. There are good facilities, including coin-operated hot showers, and it's a short drive to Whangārei for supplies. All rubbish needs to be taken with you (Uretiti Recycling Centre is on Tip Road, 4km/2.5 miles away). Nearby, you can hike the stunning Mangawhai Heads Clifftop Coastal Walk, explore limestone caves at Waipu, or visit Piroa waterfall.

THE PITCH

A large but relaxed campground set behind the sand dunes at Uretiti Beach – a white-sand surf beach with its own nudist area.

When: year-round
Amenities: unpowered tent sites, showers, cooking shelter, untreated water
Best accessed: by car
Cost: $7.50–$15
Contact: www.doc.govt.nz

16

© RUDMER ZWERVER | SHUTTERSTOCK

WILD FOREST ESTATE
NORTHLAND

17

Three rivers converge at this idyllic bush haven in the rewilded kauri-forested foothills of the Northland coast. Rustic-chic living, private outdoor bathtubs among the trees, and after-dark kiwi-spotting all guarantee a restorative return to nature at photographer Joanna Wickham's Wild Forest Estate.

Embrace the sense of off-grid seclusion at this eco-conscious retreat – but with hot baths, warmly decorated interiors, Persian rugs and linen sheets. Accommodation options include a stylish bushman's cottage, an eco-tent with bath, BBQ and hammock; a cosy off-grid container with a queen bed; a one-bedroom former school building; a tin and timber treehut; and a warehouse apartment with 'indoor camping'. It's an excellent group- or family-trip option, with space for 12 among the six varied accommodations set down dirt roads in the forest.

A visit to the incredible thousand-year-old Tāne Mahuta kauri tree in Waipoua Forest, or a stroll by the crashing surf at Aranga Beach, both within a 30-minute drive, are the perfect complements to this nature-based retreat. If you'd rather stay put, you'll be surrounded by a wealth of birdlife, and the sound of the stream flowing by – not to mention you'll have the owners' pet goat for company.

 **THE PITCH**

This lush northerly corner of the North Island offers diverse accommodation options, from off-grid glamping to a rustic cottage within 35 acres of rewilded forest.

When: year-round
Amenities: flushing toilets, hot showers, full kitchen (BYO provisions)
Best accessed: by car
Cost: $80-160
Contact: www.wildforestestate.com

© JOANNA WICKHAM PHOTOGRAPHY

PURIRI BAY (WHANGARURU NORTH HEAD) CAMPSITE

NORTHLAND

At the mouth of Whangaruru Harbour, and a long drive from Hwy 1, this campsite has a slow-down-and-forget-all-your-worries feel. The sheltered beach is ideal for swimming and activities like kayaking or stand-up paddleboarding on the usually millpond-still water. There are walks from the site which pass through native forest along ridgelines overlooking the harbour, with 360-degrees views from the surveying station (ask the camp hosts for a map). Those bays are worthy of a sea-kayaking adventure if you're game to round the headlands, and dolphins are sometimes seen off the coastline.

There are 90 unpowered tent sites here, plus enclosed and open-air (cold-water only) showers. The site can be accessed year-round, but is more popular in summer. In the off-season there are no hosts – campers need to register with the DOC website; check the campsite is open and dry before you do. There's no rubbish collection; take everything with you on departure to Ōakura or Russell, where you can also find provisions – both are a long drive away.

 **THE PITCH**

An out of the way, family-friendly campsite that spills down to a beach, with a yesteryear vibe – and kiwi-spotting opportunities.

When: late Oct–Apr.
Amenities: unpowered tent sites, cold showers, cooking shelter, drop toilets, untreated water
Best accessed: by car
Cost: $18
Contact: www.doc.govt.nz

19

PORONUI SAFARI CAMP
HAWKE'S BAY

Glamp in style and fill your days horse riding, hiking, fly fishing, bee-keeping or taking a cooking class at this high-end retreat. Yes, this is luxury glamping but with an intentional connection to the land and culture of the North Island. Resident guides share their knowledge of the region, its history and people; the onsite cooking school is very much focused on traditional Māori techniques.

Top off a big day out tramping the wilds in the steam and sauna rooms nearby, or play a round of snooker on the full-size billiard table. If all you want to do is unwind in the privacy of

your secluded safari camp, that's perfectly fine. In fact, for an extra fee, you can book a private chef for dinner or breakfast.

The camp is nestled right next to the river in the Taharua valley, not far from Lake Taupō. You'll fall asleep to the rushing water, as the river winds its way into Hawke's Bay just north of Napier.

🐾 THE PITCH

Stay in a safari tent by the Mohaka River and experience an action-packed, backcountry New Zealand on the grounds of this luxurious lodge.

When: year-round
Amenities: safari tents, food and alcohol provisions, a private kitchen tent, hot showers and flushing toilets
Best accessed: by car
Cost: $1650 per night, up to 4 guests
Contact: www.poronui.com

RIPPLES RETREAT
CENTRAL NORTH ISLAND

Both couples seeking romance and families with older children will be charmed by the attention to detail in this cosy cabin, with three 'hidden' bunk beds accommodating up to five people. After a tranquil night's sleep, you can step out onto the lawn for some wild river-swimming on your doorstep. There's limited mobile phone reception but good wi-fi, and the indoor and outdoor games available here (giant Jenga, petanque and frisbees for the kids; board games, cards and puzzles for the grown-ups) make for a fun-filled digital detox. Outdoor activities are covered with a tree swing and a double kayak to take out on the Mangaotaki River.

Mains power, hot water, an outdoor bath and a family kitchen keep it civilised at this otherwise fairly secluded cabin on a working farm. A campervan of friends (by prior arrangement) can park up next to the cabin, and plug into the outside power socket and fresh-water hoses. Provisions are available at Piopio, about 15 minutes away, as well as a top-shelf local cafe and pub. Beyond town there's a host of unique outdoor experiences to enjoy including jetboating, black-water rafting, ziplining and glowworm boat trips – all within an hour's drive. Even hitting the ski slopes is possible on an action-packed day trip.

20

© RIPPLES RETREAT

🏕 THE PITCH

There's space for up to five to stay in cosy riverside serenity on this spectacular fourth-generation family farm.

When: Oct-Jun
Amenities: BBQ, coffee machine, fridge-freezer, double outdoor bath, double shower
Best accessed: by car
Cost: from $320 per night
Contact: www.stayatripples.co.nz

MARUA RETREAT
NORTHLAND

A tasteful renovation and hotel-style additions – plush white towels and bathrobes, organic soap and a heritage fit-out – raise this simple rural cabin to something worthy of a magazine shoot. There are indoor and outdoor fireplaces, and the old-style railing fences show off the area's extremely photogenic timber. A landscaped path takes you to the separate, spacious bathroom and rustic bath pavilion with views of paddocks stippled with forest. The self-catering kitchen opens to similarly vast views of the treetops, rolling hills and big sky. Guests are treated to organic, local and seasonal produce, like free-range farm eggs, craft beer and Marua honey.

The cabin is best for solo travellers or couples – the main room has a charming four-poster king bed. It's not suitable for families at all, with potentially dangerous drops and steep hiking tracks. On the farm you'll find pockets of native bush, a haven for local birdlife (binoculars are supplied). A location near the Glenbervie Forest outside Hikurangi puts you in prime position for hiking and mountain biking – although with a setting this beautiful, you won't want to leave the cabin.

📑 THE PITCH

Rustic cabin on a spectacular, sprawling farm, with expansive sunset and sunrise views – a touch of luxury in a rural idyll.

When: year-round
Amenities: private bathroom, flush toilet, outdoor bath, BBQ, kitchen and fridge
Best accessed: by car (preferably 4WD)
Cost: from $550, 2-night minimum
Contact: www.maruaretreat.co.nz

21

© MARUA RETREAT

TOTALLY TARAWERA GLAMPING
LAKE TARAWERA, ROTORUA

22

Set on the edge of a native forest, home to the morepork (New Zealand's native owl) and the melodic tui bird, Kanuka is a tastefully decorated glamping tent, featuring elements of Māori cultural design like beautiful natural materials and traditional flax weaving. Totally Tarawera will deliver you to this remote location via water taxi and provide insights into the history and geology of the region. After being inducted into the site, you'll be left to relax. This is the perfect spot for hammock escapism, swimming, or spending an afternoon walking to the natural thermal springs at Hot Water Beach, 90 minutes away along the Tarawera Trail.

The accommodation is fully self-catered (BYO food and drinks from Rotorua), although breakfast and dinner hampers can be arranged by request. A permanent camp kitchen can handle your cooking and washing up, and the compost bathroom is nearby. A warm down duvet, extra blankets, a hot-water bottle and a heater are on hand to keep you toasty in the cooler months. Their sister site, Te Rata, is another beautifully off-the-beaten-path experience if Kanuka is booked out. Tarawera means 'a very special place', and you'll quickly come to know why.

🌀 THE PITCH

Find total seclusion at this discreet, couples-only lake-edge glampsite, accessible by water taxi or a two-to-three-hour walk.

When: Sep-Jun
Amenities: solar-lit tent, camp kitchen, fridge, composting toilet
Best accessed: by water taxi, hiking
Best accessed:
Cost: $320 for 2 people
Contact: www.totallytarawera.com

KAKAHO CAMPSITE (PUREORA FOREST PARK)
WAIKATO

Ancient rainforests of native rimu, tōtara, mataī, miro and kahikatea trees thrive in the Pureora Forest Park, west of Lake Taupō, as do the (often rare) birds that inhabit them – keep an eye out for the rare and endangered whio, or blue duck. Nestled on the Taupō side of the park, Kakaho Campsite gives you access to this regenerating forest at the centre of the North Island. It's also an excellent stopover on a mountain-bike ride through the park.

This basic campsite on a level, grassy clearing is enclosed by lush vegetation and has easy access to a shallow stream for water, or a cold-water dip (there is no shower here). A short loop walk from the site follows Kakaho stream through dense rimu forest to a viewpoint to spy Mt Pureora. The 1165m (3822ft) peak is a relatively easy hike (if you don't mind a lot of stairs) from Loop Road (alternatively, head up Toi Toi Track). At the top, you're rewarded with views across Lake Taupō, Mt Taranaki and Mt Ruapehu (on a clear day), and at lower altitudes, the forest is rich with ferns and mossy trunks.

THE PITCH

A basic campsite by a secluded stream, with easy gravel-road access to walks and mountain-biking routes through lush forests.

When: Oct-Apr
Amenities: unpowered tent sites, BBQ, untreated water, firepit, pit toilets
Best accessed: by car
Cost: $5-$10
Contact: www.doc.govt.nz

© TIA HUT

TIA HUT
MARTINBOROUGH

Country roads, windswept fields, views that just keep on going – New Zealand's farmlands, while a stark contrast with its tracts of dense wilderness, are equally tempting for a rural retreat. At Tia Hut, in the Martinborough wine region about an hour outside of Wellington, you're fully immersed in the beauty of New Zealand's bucolic landscapes, staying in a fully off-grid cabin overlooking the sheep and cattle pastures (while your hosts Emma and Owen get on with the business of farming).

Enjoy a freshly brewed coffee on the elevated wood deck, or unwind under the stars in the outdoor bathtub with views of the rolling grass hills around you. The same view is perfectly framed inside the hut by large windows. The slick modern interior belies the feeling of disconnection from the modern world, with the cabin entirely reliant on solar panels, a rain-water tank, a gas burner for cooking, and a cast iron stove for a winter fire. While there is mobile phone reception, the cabin is wi-fi-free. The town of Martinborough is nearby and offers local providores, including a gourmet grocer, to fill the small Tia Hut fridge.

🔗 THE PITCH

This off-grid minimalist cabin overlooks the South Wairarapa Valley, offering rural tranquillity on a picturesque working farm.

When: year-round
Amenities: coffee, gas stove, bathroom, fireplace
Best accessed: by car
Cost: from $250, occasional 2-night minimum
Contact: www.airbnb.com.au

NEW ZEALAND SOUTH ISLAND

The highest mountains in the land combine with the wildest beaches to provide an always-spectacular backdrop to nights in the open.

Best time: Oct–Apr
Best national parks for camping: Aoraki/Mt Cook National Park, Fiordland National Park, Paparoa National Park, Nelson Lakes National Park
Best camping trails: Routeburn Track, Inland Pack Track
National parks pass required: No, but consider a Campsite Pass for access to DOC campgrounds for a 30-night or 365-night period.
Useful contacts: www.newzealand. com; Department of Conservation (DOC) www.doc.govt.nz

Anchored by the Southern Alps, seven of New Zealand's 10 Great Walks and an encyclopaedic list of extreme and extravagant activities, the South Island is a byword for the outdoors and adventure. At the northern end, it's sunny side up around the Marlborough Sounds and Abel Tasman National Park, and from there it's an ascending scale of spectacular as you head south past tall forests, pin-sharp peaks and glaciers to the sounds that cut deep into Fiordland National Park, and the kiwis that rule the roost on Stewart Island. And all of it is just a zip of the tent, or an open hut door, away.

FREE CAMPING

Fees apply in most campgrounds, though DOC has more than two dozen free campgrounds across the South Island. Outside of national parks, there are plentiful free camping sites, although many of them are restricted to self-contained vans.

SUPPLIES

Christchurch and Queenstown have the greatest clusters of outdoors and camping stores. In Dunedin, try Bivouac Outdoor (www.bivouac.co.nz) or, if you're up north, MD Outdoors (www. mdoutdoors.co.nz) is a locally owned store in Nelson.

SAFETY

Weather changes fast, and often dramatically, on an island this skinny and this far south. Stay ahead of the fronts with the MetService's NZ Weather app. If you're camping as part of a multi-day hike, know your limits on high mountain trails, which can quickly

get rugged and challenging – prepare ahead with a range of resources and information from the New Zealand Mountain Safety Council (www.mountainsafety. org.nz).

BEST REGIONS

Marlborough & Nelson

It's easy to find a hideaway camp in a landscape this convoluted. Frayed with sounds and waterways, the island's northernmost reaches offer everything from wine to wilderness, be it along the country's most popular Great Walk or beside shores that represent 10 per cent of New Zealand's entire coastline.

West Coast

New Zealand's literal wild west sits pinched between the Southern Alps and the

The view from the Lindis Pods (left); fresh-air bathing at the French Wagon (below)

sea, providing landscapes as dynamic as the swings in weather – glaciers, driftwood-strewn beaches, the country's southernmost hot springs and towering forest. Be sure to bring the bug spray.

Canterbury

The South Island's heartland stretches from the country's tallest mountain to the shores of the Banks Peninsula, providing a wealth of landscapes around the edges of its rural plains.

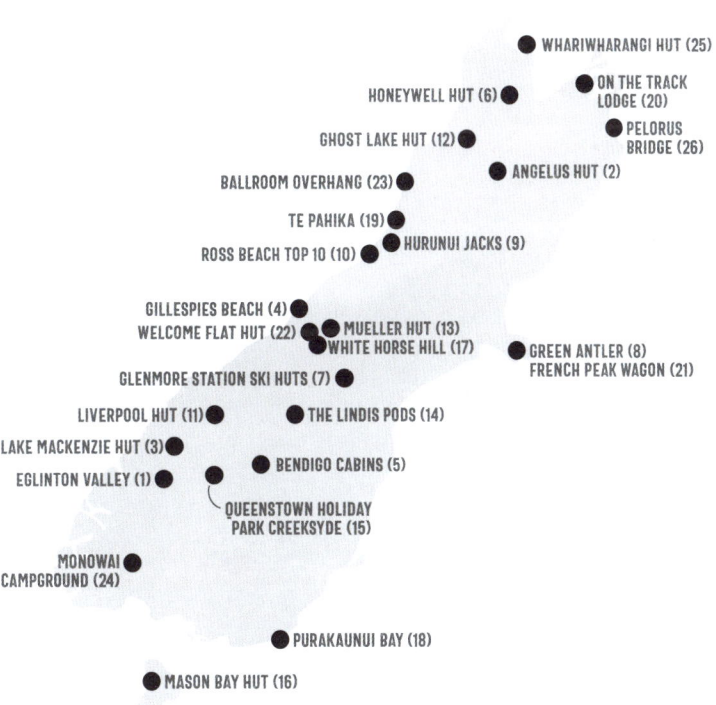

WHARIWHARANGI HUT (25)

HONEYWELL HUT (6)

ON THE TRACK LODGE (20)

PELORUS BRIDGE (26)

GHOST LAKE HUT (12)

BALLROOM OVERHANG (23)

ANGELUS HUT (2)

TE PAHIKA (19)

ROSS BEACH TOP 10 (10)

HURUNUI JACKS (9)

GILLESPIES BEACH (4)

WELCOME FLAT HUT (22)

MUELLER HUT (13)

WHITE HORSE HILL (17)

GREEN ANTLER (8)
FRENCH PEAK WAGON (21)

GLENMORE STATION SKI HUTS (7)

LIVERPOOL HUT (11)

THE LINDIS PODS (14)

LAKE MACKENZIE HUT (3)

EGLINTON VALLEY (1)

BENDIGO CABINS (5)

QUEENSTOWN HOLIDAY PARK CREEKSYDE (15)

MONOWAI CAMPGROUND (24)

PURAKAUNUI BAY (18)

MASON BAY HUT (16)

EGLINTON VALLEY
FIORDLAND NATIONAL PARK

The drive to Milford Sound from Queenstown or Te Anau is one of the most spectacular in New Zealand, but it's a long day out – more than seven hours for the return journey from Queenstown – making this pair of campsites in the Eglinton Valley an enticing option.

This spectacular, mountain-hemmed valley forms the final approach to Milford Sound before the road burrows through Homer Tunnel. One hour from Milford, the private Eglinton Valley Camp, once a sawmilling site, has unpowered sites and six self-contained cabins, all with views of the peaks that make the journey so remarkable. Five minutes' drive up the road, you'll find the DOC's tiny Upper Eglinton Campsite, which has just three first-come, first-served tent sites perched between the Milford road and the Eglinton River.

The river provides excellent fly fishing from around both campgrounds, and there's access to a range of walking tracks. The challenging Dore Pass Route crosses through the Earl Mountains to Glade House (near the start of the Milford Track), and not far away you'll find the lofty Gertrude Saddle route, with its grandstand view into Milford Sound. The trailhead for the Routeburn Track is also less than 20 minutes' drive from the camps.

🌐 THE PITCH

Break up the journey to Milford Sound in the gorgeous Eglinton Valley, home to a tiny DOC campsite and a private campground with cabins.

When: Nov–Mar
Amenities: pit toilets, river water (Upper Eglinton); hot showers, flush toilets, camp kitchen, laundry, running water (Eglinton Valley)
Best accessed: by car
Cost: sites $30, cabins from $225 Eglinton Valley; $15 per person Upper Eglinton
Contact: www. eglintonvalleycamp.nz; www.doc.govt.nz

01

ANGELUS HUT
NELSON LAKES NATIONAL PARK

There are many approaches to Angelus Hut, but none of them are easy. Perched among golden tussocks on the shores of Lake Angelus at 1650m (5413ft) among golden tussocks, it's an 800m (2625ft) ascent from the Mt Robert carpark, whichever track you take.

The most spectacular path to this most spectacular of huts, and its adjacent campsite, is the six-hour hike along Robert Ridge, a sharp crest above a sprinkling of alpine tarns, with the sort of open, expansive views you seldom encounter on walking tracks.

The 28-bunk hut, once voted the best in the country by readers of *Wilderness* magazine, has insulated walls, a spacious common area and, best of all, a large deck that overlooks the lake and the mountain backdrop. Plan on at least a two-night stay, with the hut making the perfect base for an exploration of the glacial basin overlooking the lake, the nearby Hinapouri Tarn, or the scrambling ascent of Mt Angelus.

Given its elevation, Angelus is very much a summer hut for all but very experienced trampers, with tracks blanketed in snow and at risk of avalanche from about May to October. During this time firewood is removed, and water may be turned off.

🔘 THE PITCH

Savour mountain life at its best in this popular lakeside hut, high among alpine peaks with myriad exploration options.

When: Nov-Apr
Amenities: wood heater, mattresses, untreated drinking water, pit toilets
Best accessed: by hiking
Cost: camping $15 per person, hut $30
Contact: www.doc.govt.nz

LAKE MACKENZIE HUT
FIORDLAND NATIONAL PARK

Sure, the Milford Track might wear its 'finest walk in the world' title like a historic badge of honour, but there are many who rate the Routeburn Track even higher. At the mountainous heart of this 33km (20.5-mile) Great Walk is Lake Mackenzie Hut, sprawled on the shores of its namesake lake.

This jewel of a lake is pooled in a hanging valley on the slopes of the Ailsa Mountains, 700m (2296ft) above Hollyford Valley. From its subalpine setting, the two-storey hut looks over the southern end of the lake towards Ocean Peak (around which the Routeburn Track loops as it crosses Harris Saddle) and the distant Darran Mountains,

one of the most remote and challenging ranges in the country.

It's a large hut – perhaps more a small village – with the common area downstairs and room for 30 to sleep upstairs in bunks and on a sleeping platform stretched along

THE PITCH

Poised high in the Ailsa Mountains, halfway along one of the greatest of the Great Walks, this large lakeside hut is a tramping favourite.

one wall. Attached to this main building is a separate bunkroom that sleeps another 20 people. A warden's hut completes the settlement feel.

Midway along the lake, about 10 minutes' walk from the hut, is Lake Mackenzie's small campground (nine tent sites) – one of only two along the Routeburn Track – with artificial-turf sites.

When: Nov-Apr
Amenities: gas stoves, mattresses, flush toilets, untreated water, heating
Best accessed: by hiking (only)
Cost: camping $21/$32 NZ residents/international visitors; hut $68/$102 NZ residents/international visitors
Contact: www.doc.govt.nz

GILLESPIES BEACH
WEST COAST

Do you want beach, or do you want mountain? Or do you want both, with each one as wild as the other? Gillespies Beach is the literal end of the road, with black sands strewn with driftwood, mining relics and seals, and the eight sites of its DOC campground backed by the snow-tipped Southern Alps.

Despite its beachside location, Gillespies isn't a place to come for sun, sand and surf – the ferociously chilly sea will put paid to that – but the campground has a lonely, end-of-the-world appeal enhanced by the long dirt-road approach, a range of remote walking tracks and proximity to Fox Glacier and the mountains.

The camp, near an abandoned gold-mining settlement, is the starting point for five walking trails. Short tracks lead to a miners' cemetery and the remains of gold dredges, while longer trails burrow into a tunnel carved by miners and head through gorgeous rimu forest to a fur-seal colony on remote Galway Beach. All this with views of Aoraki/Mt Cook.

The approach road to the campsite also passes by picturesque Lake Matheson, whose surface perfectly reflects the surrounding mountains, while the turnoff to the campground is in Fox Glacier, making it easy to go from beach camp to an ice heli-hike.

⊘ THE PITCH

The edge of the world feels within reach at this wild camp among seals, mining relics, glaciers, reflective lakes and views of New Zealand's highest mountains.

When: Nov–Mar
Amenities: pit toilets, cooking shelter, untreated drinking water
Best accessed: by car
Cost: $10
Contact: www.doc.govt.nz

04

BENDIGO CABINS
CROMWELL

All but concealed by the rocky outcrop on which they sit, Bendigo Cabins' twin huts offer seclusion amid the Queenstown-Wanaka bustle. Tiny houses with tiny footprints, Dunstan Cabin and Pisa Cabin peer down onto the vines of the Clutha basin and Lake Dunstan sprawled through the valley.

Standing on bedrock 100m (328ft) above the valley, the timber huts are very similar in design – elevated queen beds, wood heaters, kitchenettes, board games and books, a pair of armchairs and decks with expansive views – but each has its own character, from the lighter, brighter Pisa to the darker, moodier tones of Dunstan.

The latter's front deck steps down into a sunken outdoor bath that all but overhangs the valley, while Pisa's outdoor tub is set among the rocks that armour-plate the outcrop. It's completely immersed in nature, and yet it's an easy squeeze through the Kawarau Gorge to Queenstown and the wineries of the Gibbston Valley, and an even shorter journey into Wanaka. Even closer at hand, the brilliantly engineered Lake Dunstan Trail awaits, the newest of the Great Cycles of New Zealand, with boardwalks clipped to the cliffs that jut out over the lake.

🔖 THE PITCH

Two tiny, well-hidden hilltop cabins overlooking the Clutha basin, within an hour's drive of Queenstown and Wanaka.

When: year-round
Amenities: wood heater, hot shower, flush toilet, outdoor bath, running water
Best accessed: by car
Cost: from $219 for 2pp for 2 nights
Contact: www.bendigocabins.co.nz

05

HONEYWELL HUT
MOTUEKA

06

Not everything is as it appears at Honeywell Hut. Looking every bit a weathered backcountry musterers' hut, it actually began life as a changeroom at a Nelson swimming pool before being trucked onto Baton Run Farm and completely refurbished – even much of the wood in the structure was replaced with timber from this beef and sheep farm in the Baton Valley.

The one-room, off-grid hut, sleeping up to five people, is set in pasture a few metres from the Baton River. Pieced together with recycled materials, including a bunk bed made from the old cattle yard, and lined with macrocarpa milled from a neighbouring property, it's suitably equipped in rustic farm style. An old farmhouse stove sits outside, heating the water for the open-air bathtub, and a quirky firepit with concrete couch and tractor-saddle seats is a short walk away on the riverbank.

Nelson and Abel Tasman National Park are an hour's drive away, but there are also walking trails up into the hills on the property, providing views to Kahurangi National Park.

Baton Run also has two other huts – the equally rustic Shepherd's Rest, and Pedallers' Rest, a two-room hut aimed at cyclists on the Great Taste Trail, which runs almost right past the door.

🌱 🥾 THE PITCH

A public-pool changeroom turned farm hut with distinctly local touches, that's both remote and in easy reach of Nelson and Abel Tasman and Kahurangi national parks.

When: Oct–Apr
Amenities: outdoor bath, hot shower, running water, flush toilet, wood heater, full kitchen
Best accessed: by car
Cost: from $230
Contact: www.batonvalley.co.nz/honeywell-hut

GLENMORE STATION SKI HUTS
LAKE TEKAPO

Wedged between Lake Tekapo and the Tasman Valley, the Gamack Range provides some of New Zealand's best backcountry ski-touring terrain, aided by the presence of six private huts on Glenmore Station, 190 sq km (73 sq miles) of merino, cattle and deer land.

Three of the huts – Waterfall Hut, Tin Hut and Memorial Hut – sit nestled into the Cass Valley, with O'Leary's Hut, Lady Emily Hut and Falcon's Nest commanding higher positions, ranging from 1500m (4921ft) to 1800m (5905ft) above sea level on the slopes of the range. They are a boon for skiers, creating a welcome base from which to explore. Clear nights bring the promise of skies strewn with stars, with Glenmore Station encompassed by the Aoraki Mackenzie International Dark Sky Reserve.

Each hut sleeps between eight and 10 people, and most have been purpose-built for skiers – Waterfall Hut (with said waterfall pouring down behind it) is an original musterers' hut, creating a bit of extra farm authenticity. Stay at Lady Emily or Falcon's Nest and there's the additional option of skiing out to the Bad Decision Hut, a ridgetop, bivvy-style shelter-cum-bar. Organise ahead and a bottle of whisky will be delivered by helicopter and waiting for you.

07

💬 THE PITCH

Six private huts dotted across the Gamack Range behind Lake Tekapo provide ski tourers with an overnight base, and perhaps even a sneaky whisky.

When: July–mid-Oct
Amenities: wood heaters, gas stoves, mattresses, solar lighting, toilets
Best accessed: by car, skiing
Cost: from $40
Contact: www.glenmorestation.co.nz

GREEN ANTLER
LITTLE RIVER

Glamping often seems to be about cutesy couples and loved-up nights away, but Green Antler puts the fam into glam with a safari tent that sleeps up to five people. Set on a sprawling Banks Peninsula deer farm, the standalone tent has a spacious lounge area wrapped around a wood-burning stove. A deer-antler lampshade provides a local touch, and a queen bed sits beyond the lounge, in a space that can be zipped down to form a separate bedroom. A second room contains bunk beds for the brood. On the tent porch is a small, covered kitchen area with BBQ for nights when it's easier to feed the clan in than out. The bathroom stands a few metres behind the tent.

Out front, commanding pride of place, is a wood-fired hot tub (equipped with plush bathrobes) looking out over the peninsula's many ridges and pockets of forest. The tub takes a few hours to heat – notify the owners and they'll have it warmed and ready for your arrival.

It's a 10-minute drive into Little River, one end of the 60km (37-mile) Little River Rail Trail that runs all the way back to Christchurch. *Tres chic* Akaroa, New Zealand's Frenchest town, is another 30 minutes by road.

 THE PITCH

Wandering deer, board games, and total privacy less than an hour's drive from Christchurch. Welcome to glamping, family-style.

When: year-round (couples only Jun-Sep)
Amenities: hot showers, composting toilet, outdoor hot tub, outdoor kitchen, BBQ
Best accessed: by car
Cost: from $680 for 2 nights
Contact: www.canopy camping.co.nz/green-antler

Wildlife

New Zealand might be more famous for the animals it doesn't have – no native land mammals, no snakes – than the ones it does, but the South Island delivers plenty of wildlife encounters.

In this country where even the people brand themselves after the national bird – kiwis – wildlife watching is a big deal, and nowhere more so than the wilderness-rich South Island. A wealth of birdlife populates the forests, from tiny but curious South Island robins to the heavyweight kererū, the native pigeon that can tip the scales at almost 1kg (2lb). The mischievous kea is the world's only alpine parrot (leave your backpack unguarded at your peril) and if you're tramping along the Greenstone Track, you might now even sight a takahē, the colourful, flightless bird once thought extinct, with 18 of the birds released onto Greenstone Station in August 2023.

The coast is the South Island's other great treasury of wildlife, home to fur seals, sea lions, Hector's dolphins, yellow-eyed penguins, sperm whales, orcas and humpback whales.

KIWIS RULE

New Zealand's third island, Stewart Island, has a population of just 400 people, but it's also home to an estimated 13,000 southern brown kiwis – more than 30 of the birds for every human. From Oban, nightly tours head to a long spit at the mouth of Paterson Inlet where kiwi sightings are common – it was here that Sir David Attenborough filmed kiwis for *The Life of Birds* documentary.

BEYOND MAMMALS

New Zealand might not have native land mammals, but there's plenty else that calls these islands home. The country is considered the world's seabird capital, with around 25 per cent of the planet's seabird species breeding here. There are 51 native freshwater fish species, three native frogs, and reptiles are headlined by the prehistoric tuatara, which once roamed among the dinosaurs. And don't let anyone tell you that the country has no native mammals at all, it has two: the long-tailed bat and the short-tailed bat.

Yellow-eyed penguins on Otago Peninsula (top); a fur seal on the Kaikoura coast (above); a Hector's dolphin leaping (left) and a kiwi foraging (inset)

FIVE TO TRY

Ulva Island/Te Wharawhara
Predator-free, unlogged bird sanctuary with a trail at its western end.

Kaikoura
Take in New Zealand fur seals, plus sperm whales, humpbacks and orcas on seasonal cruises.

Catlins
View New Zealand sea lions from camp at Purakaunui Bay, and glimpse Hector's dolphins at Curio Bay.

Otago Peninsula
Gawp at the wingspan of the northern royal albatross and encounter yellow-eyed penguins.

Oamaru
A grandstand view of blue penguins marching ashore after a hard day of ocean work.

09

HURUNUI JACKS

KOKATAHI

With a literary setting and one of New Zealand's Great Rides rolling right past its glampsite, the lone luxury tent at Hurunui Jacks is perfectly positioned for a West Coast escape. Enclosed by rainforest, but just a short drive from the cafes and driftwood-strewn beaches of Hokitika, the roomy tent is all about warmth in the wet wilderness. A wood stove and goose-down duvet ensure things are cosy inside, while the deck steps down to a firepit and robes hang beside the bed for trips to the outdoor bath set into the ferny bush.

To this, add a leather couch, small dining table, Adirondack chairs on the deck, a separate kitchen and bathroom with views into the forest, and the Kaniere River – the setting for the Chinese camp in Eleanor Catton's 2013 Booker Prize–winning novel *The Luminaries* – just a few metres away, down a short path. A private bench on the riverbank is the prime morning-coffee spot.

Cyclists, in particular, will find Hurunui Jacks a restorative stop on the 133km (83-mile) West Coast Wilderness Trail, which runs right through the property as it follows a gold-rush-era water race towards large Lake Kaniere, pooled at the foot of the Southern Alps.

THE PITCH

A trailside glamping tent wrapped in ancient rainforest – perfect as a cycling stop, or a cosy base from which to explore the West Coast.

When: Sep-May
Amenities: kitchen, hot shower, flush toilet, outdoor bath, firepit
Best accessed: by cycling, car
Cost: from $250
Contact: www.hurunuijacks.co.nz

© DAN KERINS

ROSS BEACH TOP 10
ROSS

Caravan parks often have a sameness to them, with only the setting varying, but Ross Beach Top 10 has much more than the wild West-Coast beach at its edge.

The main point of difference is in the 11 cabins dotted through the small park, fashioned from shipping containers. Pods come with built-in kitchen and bathroom (with plush bath robes), queen bed and decks with BBQs, while the Sunset Studio Apartment Pod delivers an amazing beachfront location. The Whanau Sleeper Pod sleeps four, making it a good choice for families.

Tent campers have the choice of three grassed areas, campervans can set up right at the beach's edge, and four geodesic glamping domes were added in 2023, each with a kitchen and a freestanding bath at the foot of the bed, peering through the dome's private windows to ocean sunsets.

In the evenings, head to the 'chill-out zone', equipped with bean bags and chairs around a stone fireplace and open firepits, with wood-fired pizzas, antipasti and drinks available November to May.

The four- to five-day West Coast Wilderness Trail, one of New Zealand's 23 Great Rides, begins (or ends) right next to the park, which offers car storage while you're riding.

THE PITCH

The beach couldn't get any closer at this caravan park that puts a twist on standard cabins, with converted shipping containers and brand-new geodesic domes.

When: Oct–Apr
Amenities: hot showers, flush toilets, wi-fi, running water, powered and unpowered sites
Best accessed: by car
Cost: sites from $56, pods from $150 (2-night minimum), domes from $325 (2-night minimum)
Contact: www.top10.co.nz/park/ross-beach-top-10

10

© ROSS BEACH TOP 10

LIVERPOOL HUT
MT ASPIRING NATIONAL PARK

There are few huts in New Zealand that can equal the views of Liverpool Hut, and even fewer that have toilets with panoramas this magnificent. A short walk downhill from the hut, perched on the point of a spur, Liverpool's small loo-with-a-view stares straight through the Matukituki Valley 500m (1600ft) below, and out to the high peaks that frame the valley.

Sleeping just 10 people, Liverpool Hut seems like a cuter version of the ever-popular Mueller Hut, with the bright-red shelter and its equally garish toilet standing among tussocks 1100m (3600ft) above sea level.

Getting here requires a walk of up to seven hours, setting out, like so many of this park's tracks, from the Raspberry Creek carpark, and following the flat course of the Matukituki Valley – the main street of sorts for Mt Aspiring National Park – to the popular Aspiring Hut. After this point, the track ramps up steeply, requiring the use of hands and feet over ladder-like tree roots and rocks to reach lofty Liverpool.

The hut is cosy and communal, as you'd expect of a small hut in the mountains, but the greatest attractions are outside, in the changing mountain light of evening and morning, and in that outrageously positioned dunny.

● THE PITCH

A high mountain hut that's short on space – just 10 bunks – but long on alpine vistas. Have you ever seen a toilet with a view this good?

When: Nov–Apr
Amenities: mattresses, pit toilet, untreated drinking water
Best accessed: by hiking
Cost: $25
Contact: www.doc.govt.nz

11

12

GHOST LAKE HUT
OLD GHOST ROAD

Approach the near-highest point of the Old Ghost Road, somewhere near its midway point, and you'll come to a hut perched atop a cliff above Ghost Lake. For mountain bikers on this shared-use 85km (53 miles) former gold-mining route, Ghost Lake Hut makes the obvious overnight stop – plus, it's the best positioned of the Old Ghost Road's six huts.

The hut has three sleeping compartments (18 bunks total) radiating off a central kitchen and living space. Outside, there is one tent site and two summer sleepouts, each containing two double mattresses, affording an extra level of privacy for those craving more solitude.

The hut's position means that views abound in every direction, including down to the small Ghost Lake, ringed with beech forest. Also visible is the coiling track as it descends from the hut and climbs over Skyline Ridge – a virtual mud map of the next morning's journey.

The approach to Ghost Lake Hut is spectacular, climbing through rainforest, crossing earthquake slips and edging above sheer drops. An indication of the route's ruggedness and brutality is the fact that the miners never finished construction of the Old Ghost Road. It took 21st-century track-building techniques to finally complete it.

THE PITCH

A clifftop hut with private sleepouts and sweeping views, at the heart of the wild and wonderful Old Ghost Road.

When: Sep-May
Amenities: gas stove, cooking equipment and crockery, mattresses, composting toilet, rainwater tank
Best accessed: by mountain biking, hiking
Cost: $50/160/360 tent/hut/sleepout through-user fee for up to 4 nights
Contact: www.oldghostroad.org.nz

MUELLER HUT
AORAKI/MT COOK NATIONAL PARK

Mueller Hut is Aoraki/Mt Cook National Park's ultimate night out. Perched on stilts atop the Sealy Range, 1800m (5905ft) above sea level, the little red shed affords grandstand views over glaciers and snowcapped mountains, crowned by Aoraki/Mt Cook.

The only way to Mueller is on foot, and it's not an easy trek. From the Hooker Valley, the route ascends more than 1000m (3280ft) in just 5km (3 miles), passing the spectacular Sealy Tarns along the way. Most hikers make the walk to and from Mueller Hut in a day, but those who stay the night get the greatest reward. The hut's setting is truly high alpine – as close as you can get to mountaineering without an ice axe and crampons – and the hut is roomy, with decks and benches for absorbing the high-mountain scenery.

Rising behind the hut, Mt Ollivier makes for an easy side trip and a brush with mountaineering fame – this was the first mountain Sir Edmund Hillary ever climbed. The current structure is the hut's fifth incarnation – an indication of its precarious position. The first Mueller Hut was built in 1914–15, replaced in 1950 by a new hut that was promptly smashed by an avalanche. The current shelter was built in 2003.

THE PITCH

Experience the high life in this iconic 28-bunk hut sitting eyeball-to-eyeball with Aoraki/Mt Cook, at the end of one of New Zealand's best short hikes.

When: Nov-Apr
Amenities: gas stoves, solar lighting, mattresses, drop toilet, untreated drinking water
Best accessed: by hiking (only)
Cost: $45
Contact: www.doc.govt.nz

13

© AUTUMN SKY PHOTOGRAPHY | SHUTTERSTOCK

14

THE LINDIS PODS
===================

THE LINDIS PODS

AHURIRI VALLEY

High-country and high-end, this trio of pods are the ultimate place to reflect, largely due to the mirrored floor-to-ceiling glass that wraps around three sides of each one. Just 18 sq metres (194 sq ft) each, but big on views and light, the pods sit above the Ahuriri Valley on Ben Avon Station, a high-country sheep and cattle property.

The double-glazed, one-way glass provides 180-degree views of the valley and the often-snowcapped ranges that enclose it, and transform into stellar stargazing screens at night. Size is little impediment to comfort, with each pod fitting a king bed and a stylish ensuite for showers with a mountain view. Outside, decks and hot tubs provide the same views, with bonus fresh air.

The pods are remote, but it's still just a short walk to the Lindis' magnificent lodge, and all meals are included in the price of a stay (as are minibar raids). Ben Avon Station backs onto the Ahuriri Conservation Park, and the Ahuriri River is an excellent trout fishery. E-bikes are on hand to tour the station, and guided horse rides and helicopter trips or private chef-prepared picnics can be arranged. Or you can just sit and admire the mountains reflected in the windows of your pod.

🌀 THE PITCH

Thirty minutes' drive from Omarama, these three glass-walled luxury pods invite in sweeping views over a classic high-country river and surrounding mountains.

When: year-round
Amenities: hot showers, outdoor hot tub, flush toilets, running water, all meals
Best accessed: by car
Cost: from $2345
Contact: www.thelindisgroup.com

QUEENSTOWN HOLIDAY PARK CREEKSYDE
QUEENSTOWN

It isn't just the five-minute walk into the city centre that elevates this quirky holiday park. In 2004, Creeksyde, which has been operated by the same family for nearly 40 years, became the world's first holiday park to be certified under EarthCheck, recognising the park's ongoing commitment to sustainability. This began with the repurposing of the original buildings in the 1980s and early adaptation of solar water heating, and extends to the reuse of bath towels as cleaning cloths, compost bins for guest food scraps, chemical-free cleaning and gardening, and the collection of used toilet rolls to serve as fire starters.

Once a plant nursery, the park on the bank of Horne Creek has oodles of character. The bathrooms are disguised as a hop kiln and painted in trompe l'oeil scenes, from a staircase descending to a basement in the gents, to a Skyline Gondola scene in one of the ladies' showers. The various BBQ areas include a thatched hut and a plumbing-themed 'Tapas Pavilion', and there's locked bike (and ski) storage and bike wash-down stations, as well as a private sauna and spa.

With powered sites only, the park is geared more towards campervans than tents, and it has a range of cabins and units.

15

● THE PITCH

Close to the Queenstown action, with impeccable eco-credentials, this holiday park is as original as it is organic.

When: year-round
Amenities: hot showers, flush toilets, wi-fi, running water, camp kitchens
Best accessed: by car, bus
Cost: campsites from $61, units from $139
Contact: www.camp.co.nz

© REBECCA IGGO–BROWNE

MASON BAY HUT
RAKIURA NATIONAL PARK

At night in this remote corner of Stewart Island, you'll almost certainly encounter more kiwis than people. Mason Bay is said to have the world's highest population of the hard-to-spot bird, making it one of the best places for a sighting. Or just lie back and listen to their calls.

The 20-bunk Mason Bay Hut sits at the southern end of a remote, 14km (8.7-mile) West-Coast beach backed by one of the Southern Hemisphere's most extensive inland dune systems. Once primitive, the green timber hut was given a makeover in 2006 and now contains three bunkrooms and a large kitchen and dining area.

It's a muddy two- to three-day walk across the island to Mason Bay from Oban along the 125km (77.7-mile) North West Circuit, or Stewart Island Flights can deliver you by air with a low-tide beach landing. Kiwis are the star, but not the only attraction of a stay at the hut. Scale the 156m (512ft) Big Sandhill, the bay's tallest dune, just a short stroll away, and head about 20 minutes inland to find historic Island Hill Homestead, a farmhouse and woolshed established in 1880, complete with the lingering scent of lanolin and shearing tallies still pencilled onto the walls.

● THE PITCH

Find yourself surrounded by the elusive kiwi at this isolated hut tucked into the dunes of Stewart Island's North West Circuit.

When: Nov-Apr
Amenities: heating, mattresses, pit toilet, untreated drinking water
Best accessed: by hiking, plane
Cost: $10
Contact: www.doc.govt.nz

16

© VICTOR SUAREZ NARANJO | SHUTTERSTOCK

Great Walks

The South Island is the heartland of New Zealand's prized collection of Great Walks, featuring seven of the 10 trails, from the fabulously famous to the newest member of the family.

In 1908, Great Britain's *Spectator* magazine was unequivocal in declaring the Milford Track the 'finest walk in the world', helping draw global attention to the sublime pleasures of hiking on the South Island. Today, the Milford Track is just one of seven Great Walks across the South Island, and the surest way to start an evening debate in a hut is to declare a favourite walk – some still love the Milford, while others laud the nearby Routeburn Track.

A once static collection of routes, the Great Walks network has, in recent years, become more dynamic and even more weighted towards the South Island. In 2020, the West Coast's Paparoa Track became the first new addition to the Great Walks in 25 years, and the far-southern Hump Ridge Track is set to become the eleventh Great Walk as early as 2024.

PLANNING

The most popular Great Walks, such as the Abel Tasman Coast Track and the Milford Track, book out fast (the system crashed in 2023 due to the volume of people trying to secure spots on the Milford Track when bookings opened) so forward planning is prudent. Bookings for the Great Walks season, which runs from the end of October to the end of April, typically open in April. It's also possible to do some of the Great Walks without reservations out of season. Since 2018, international visitors have paid up to 50 per cent more for Great Walks than New Zealand residents.

HUTS & CAMPS

In the pantheon of Department of Conservation (DOC) huts, Great Walk huts are the most comfortable and complete in the country, containing mattresses and heating, as well as often having solar lighting and gas rings for cooking. Campsites are typically near to huts, allowing for shared used of toilets and water tanks. There's no camping on the Milford and Paparoa tracks.

Hiking the Routeburn Track (top) over Harris Saddle (above); following the coastline of Abel Tasman National Park (left); a kaka parrot (inset)

FIVE TO TRY

Routeburn Track *(33km/21 miles one way; 2-4 days)* Famous alpine crossing with a day above the treeline on Harris Saddle.

Heaphy Track *(78km/49 miles one way; 4-6 days)* Trek from native forest and rolling tussocks to wild West Coast shores.

Abel Tasman Coast Track *(60km/37 miles one way; 5 days)* Follow the beach-lined shores of sunny Abel Tasman National Park.

Kepler Track *(60km/37 miles loop; 3-4 days)* Well-designed loop with grandstand views over Fiordland's peaks and lakes.

Rakiura Track *(32km/20 miles loop; 3 days)* Rise above the mud on beautiful bays and coast. Keep an ear out for kiwis.

WHITE HORSE HILL CAMP
AORAKI/MT COOK NATIONAL PARK

When location is everything, White Horse Hill Campground is indeed the whole shebang. Just 2km (1.2 miles) north of Mt Cook Village, the large, open DOC campground sits in the middle of an alpine valley, with high ridges rising to either side like bookends and the slopes of Mt Sefton towering overhead. And it's only a short stroll to views (weather permitting) of Aoraki/Mt Cook.

The campground is prime real estate for hikers, with two of the national park's most popular tracks setting out from its edges. The Hooker Valley Track offers more bang for your buck than almost any other trail on the South Island, following the Hooker River for five easy kilometres to an iceberg-filled lake at the base of Aoraki/Mt Cook. The Mueller Hut Route also departs from White Horse Hill. There are rewards for even less effort, with Freda's Rock (monument to Freda Du Faur, the first woman to climb Aoraki/Mt Cook) just a few hundred metres down the Hooker Valley Track, and Kea Point, overlooking the silty lake at the end of Mueller Glacier, just down the Mueller Hut Route.

White Horse Hill takes campervans and caravans and has plenty of space among the alpine grasses for tents.

🔘 THE PITCH

A 60-site, unpowered campground at the heart of New Zealand's loftiest national park, with easy access to two of the country's most prized walking trails.

When: Sep-May
Amenities: hot showers and flush toilets (both closed in winter), running water, picnic tables, cooking shelter
Best accessed: by car
Cost: from $15
Contact: www.doc.govt.nz

PURAKAUNUI BAY

OTAGO

Purakaunui Bay is nature showing off. Its wide beach provides the only break in a long line of cliffs that have had their moment of Hollywood stardom, moonlighting as the castle of Cair Paravel in the *Chronicles of Narnia* films.

Located at the beach's southern end, the 40-site DOC campground is one of the coastal treasures of the already wild and spectacular Catlins. Known to surfers as PK Bay, Purakaunui has a consistent winter beach break, drawing board riders into its chilly waters. Among the rocky shores at both ends of the beach, there's plenty of opportunity to wander and explore tidal rockpools, or spot sea lions.

Inland, it's a 15-minute drive from the campground to one of Otago's most famous waterfalls, the 20m (65ft), three-tier Purakaunui Falls, and 30 minutes to Jacks Blowhole, where heavy swells push through 200m (650ft) of tunnels to blast out of a deep cavity.

Purakaunui Bay is a good base from which to further explore the Catlins. There's only one town of note – Owaka, 17km (10.5 miles) from the campground – but a world of wild wonders. Seek out Curio Bay, with its fossilised forest and Hector's dolphins playing in the surf. It's also a nesting site for Otago's endearing but endangered yellow-eyed penguin.

● THE PITCH

One of the prettiest spots along one of New Zealand's wildest stretches of coastline, with tent and campervan sites beside surf, sand and soaring cliffs.

When: year-round
Amenities: pit toilets, fireplaces, untreated drinking water
Best accessed: by car
Cost: $10
Contact: www.doc.govt.nz

18

TE PAHIKA
MARSDEN

19

Stays in converted railway carriages aren't entirely unusual on the South Island – they can also be found at Nydia Bay, Kaikoura, Hapuku and Raglan – but none of them have a backstory quite as evocative as Te Pahika. The name translates as 'the runaway', referencing the 1963 day when this guard's van, built in 1907, was part of a train that had 38 wagons disconnect and hurtle downhill, crashing by Lake Brunner.

Today, the refurbished runaway carriage sits at the edge of native West-Coast bush, with a queen bed taking up one end of the timber-lined carriage, a small seating area just inside the door and a dressing room and luggage storage space hidden away at the other end.

The rail theme extends to a cute purpose-built platform and 'station' right beside the carriage, housing a kitchen and bathroom. Step into the bush behind the station and you'll find an open-air bath beneath a timber shelter.

Te Pahika is 10 minutes' drive from Greymouth, with easy access to the West Coast Wilderness Trail. Or you could continue the railway theme – Shantytown Heritage Park, with its small collection of historic trains, is just a few minutes' drive from the property.

🔘 THE PITCH

A one-time runaway train converted into accommodation for two turns a West-Coast stay into a brush with history.

When: Oct–Apr
Amenities: hot shower, flush toilets, running water, kitchen, wi-fi, USB charging ports
Best accessed: by car
Cost: from $155
Contact: www.airbnb.co.nz

© TE PAHIKA

ON THE TRACK LODGE
NYDIA BAY

Not all track accommodation is created equal. At the midpoint of the two-day Nydia Track – the Queen Charlotte Track's quieter sibling – you'll find a couple of curious options at waterside On the Track Lodge.

Set on Nydia Bay, a four- to five-hour walk from either trailhead, the lodge features (among other possibilities) a seaside cabin, a Mongolian-style yurt and a historic train carriage. The 1930s carriage is divided into two rooms, each featuring a double bed and ensuite, while the yurt has five single beds for small groups. The compact cabin has the best real estate, being in earshot of the sea, with water views from the mezzanine bedroom. All rooms grant access to the lodge's outdoor hot tub.

The lodge is strung between two saddles along the shared-use Nydia Track. There's no road access, but it can be reached by boat – either water taxi or the mail boat that's been delivering post to Nydia Bay for more than a century.

The Nydia Track is one of the few overnight tramps that allows dogs (a free permit from DOC is required). Suitably, On the Track Lodge also welcomes pooches – the lodge has separate dog accommodation space.

All stays include full board, with a packed lunch for the track.

20

© ON THE TRACK LODGE

🗨 THE PITCH

Break up a two-day tramp on the Nydia Track with a seaside yurt or train-carriage stay and a restorative hot tub.

When: Oct–Apr
Amenities: hot showers, all meals, free kayak and SUP use, outdoor hot tub
Best accessed: by hiking, mountain biking, boat
Cost: cabin $250, carriage $400, yurt $750
Contact: www.onthetracklodge.nz

FRENCH PEAK WAGON
BANKS PENINSULA

Picture a cedar bathtub set on the edge of a verdant vineyard, steps from a wagon that seems to have rolled straight from a Steinbeck novel or Hollywood Western. Beneath the hooped canvas of the wagon is a quirky space featuring a queen bed, wood heater, kitchenette, shower and toilet. Giddy up!

French Peak Wagon sits on the fringes of the Banks Peninsula's oldest vineyard, looking out over French Farm Bay, which reaches deep into the peninsula. Take in bay views from every angle – lounging in bed, soaking in the wood-fired bath, or with a coffee at the elevated breakfast bar at the rear of the wagon. A furnished wooden front deck provides another spot for relaxation.

From the wagon, it's a 10-minute walk down to the bay. Akaroa is a 20-minute drive away, and it's barely an hour back into the big smoke of Christchurch. At some point, the French Peak cellar door will likely beckon – grab a bottle of pinot noir or chardonnay, fill up the tub and settle in for a starry evening.

THE PITCH

Circle your wagons for a winery stay with a difference – a Western-style covered wagon offering coastal views and star-filled nights in the tub among the vines.

When: year-round
Amenities: hot shower, kitchenette, wi-fi, running water, compost toilet
Best accessed: by car
Cost: from $395
Contact: www.frenchpeak.co.nz

© FRENCH PEAK WAGON

22

WELCOME FLAT HUT

WESTLAND TAI POUTINI NATIONAL PARK

From just south of Fox Glacier, the Copland Track sets out from SH6, passing through tōtara and rimu forest and open river flats as it follows the Karangarua and Copland rivers for 18km (11 miles) to Welcome Flat Hut, beneath the nose of Aoraki/Mt Cook.

This large timber hut has 31 bunks spread across four rooms on the hut's upper level, above the common area. Also within the hut is the Sierra Room, once the warden's quarters but now able to be booked as a private room sleeping up to four people – it has its own wood stove, shower, lighting and gas cooker (note that the rest of the hut has no stoves).

Camping is also available around the hut.

Welcome Flat Hut's finest feature, however, is delivered by nature. A short stroll away, New Zealand's southernmost thermal springs emerge from the earth, bubbling up at around 60°C (140°F) and flowing through a chain of three shallow pools. The first pool can be as hot as 55°C (131°F), so most people prefer the second. Settling into its warm embrace, with views of the Sierra Range for company, is very 'welcome' indeed after the seven-hour hike to get here.

🔵 THE PITCH

This backcountry DOC hut is worth the long walk in, with the twin treats of a private room and a natural hot spring, perfect for soaking weary muscles.

When: Nov-Apr
Amenities: heater, mattresses, pit toilets, untreated drinking water
Best accessed: by hiking
Cost: $30, Sierra Room $132
Contact: www.doc.govt.nz

BALLROOM OVERHANG
PAPAROA NATIONAL PARK

23

Life as a cave-dweller has never looked as enticing as at Ballroom Overhang. There is indeed a ballroom-like grandeur and scale to the overhang, set beneath leaning limestone cliffs, though there's no record of it ever actually being used as a ballroom. Set on the bank of the Fox River, 500m (1640ft) upstream from its confluence with Dilemma Creek, the cave is most definitely a favourite among hikers and campers.

The overhang is 10m (33ft) high in places and 30m (98ft) deep, running like a groove for 100m (328ft) along the cliffs, making a stay here almost like pitching in a cave. The cliffs provide ample cover from any rain and other adverse weather, while the ceiling of the overhang is blanketed in ferns, vines, grasses and shrubs, creating the ultimate hanging garden.

Other than a few benches cobbled together by previous campers, the Ballroom has almost no facilities and can only be reached on foot. It's found along the Inland Pack Track, an advanced two-day tramping route that requires good river-crossing skills. It's just a 6km (3.7-mile) walk from the mouth of the Fox River, or 19km (12 miles) from the track's southern end near Punakaiki.

⊛ THE PITCH

It doesn't get much closer to nature than this palatial cavern campsite in verdant Paparoa National Park.

When: Dec–Mar
Amenities: pit toilet, untreated river water
Best accessed: by hiking
Cost: free
Contact: www.doc.govt.nz

MONOWAI CAMPGROUND
FIORDLAND NATIONAL PARK

With the distracting presence of Milford Sound, it's easy to forget that Fiordland National Park stretches away another 200km (124 miles) to the south. It's here, closer to the Southern Ocean than the famous fiords, that you'll find the quiet waters of Lake Monowai, offering tranquil walking tracks and a DOC campground as remote as any on the South Island.

Simplicity is the appeal here, with five tent sites at the northern tip of the sickle-shaped glacial lake ringed with tree stumps from the time, a century ago, when the water level was raised to power one of the country's first hydro stations.

Surrounded by mountains and superb native bush, the campground doubles as the trailhead to other lake scenes and further-flung huts. The Green Lake Track climbs through a pass to the beauty spot of Green Lake, with Green Lake Hut in a lakeside clearing on its shores, looking over the lake and through a break in the mountains to distant, snowy peaks. Sticking to Lake Monowai's shores, a second trail heads south to the thumb-like Rodger Inlet, site of two DOC huts: the historic, two-bunk Rodger Inlet Hut, which is like a precursor to modern tiny houses, and the newer six-bunk Rodger Inlet Hut.

THE PITCH

As far from Milford Sound as Fiordland can get, this five-site lakeside DOC campground reveals a very different side to the park.

When: Nov-Apr
Amenities: pit toilets, untreated drinking water
Best accessed: by car
Cost: free
Contact: www.doc.govt.nz

24

© PAUL GRACE PHOTOGRAPHY SOMERSHAM | GETTY IMAGES

WHARIWHARANGI HUT
ABEL TASMAN NATIONAL PARK

Whariwharangi Hut is among the most unusual of New Zealand's Great Walk huts. Located near the northernmost point of the Abel Tasman Coast Track, the two-storey hut began life as a farmhouse in the 1890s, before being abandoned roughly three decades later. It would be another 50 years before the building was restored as a hut for trampers on the country's most popular Great Walk.

Inevitably, that history provides an extra layer of character to the 20-bunk timber hut in a grassy clearing behind a typically spectacular Abel Tasman beach. It's the most beautiful, and also the quietest, of the Coast Track's huts – the eye in the storm of sorts, as the track's 40,000-plus annual trampers march past after a final night spent at Anapai Bay or uber-popular Tōtaranui.

The hut, which also makes an easy overnight hiking destination from Tōtaranui or Wainui, retains some of the charm of a farm building, while also being at the heart of the Coast Track's most impressive section. Venture out to nearby Separation Point for both fur seals and the walk's most dramatic outlook, or wander on to Mutton Cove and Anapai Bay to find some of the park's finest beaches.

25

THE PITCH

The Abel Tasman Coast Track's quietest hut is also its most enticing, rich in history and with easy access to the track's finest stretch.

When: Sep-Nov, Feb-May
Amenities: wood heater, mattresses, flush toilets, untreated drinking water
Best accessed: by hiking (only)
Cost: NZ residents $42, international visitors $56
Contact: www.doc.govt.nz

PELORUS BRIDGE
MARLBOROUGH

26

Stand on the bank of the Te Hoiere/Pelorus River at Pelorus Bridge and you half expect to see barrels bobbing past – it was here that hobbits fled through raging rapids in the famous barrel scene of Peter Jackson's *The Hobbit: the Desolation of Smaug.*

The beautiful DOC campground extends across both banks of the river, right beside the bridge and beneath tall rimu, kahikatea and beech trees that form one of the last river-flat forests in Marlborough.

Combined, Pelorus Bridge's five camping areas have 54 sites, including 14 powered sites on the northern bank. Among the unpowered sites on Kahikatea Flat is a laundry, cafe (closed at the time of writing, with a tender out for new operators) and communal kitchen – the latter's deck has such brilliant views over the river that it's worth staying in to cook.

It's possible to swim, kayak and fish in the river, and the campground is also the trailhead for several day walks, be it a 30-minute stroll on the Totara Walk or a four-hour ascent of Trig K. The three- to four-day Pelorus Track begins just 13km (8 miles) down the road, offering the chance to hike beside the river's deep-green pools almost to the edge of Nelson.

THE PITCH

Barrel into this forest-wrapped, riverside DOC campground in a Hollywood setting, with access to multiple walking trails and swimming holes.

When: Nov–Apr
Amenities: hot showers, flush toilets, gas stoves, laundry, powered sites, running water
Best accessed: by car
Cost: from $20
Contact: www.doc.govt.nz

INDEX

AUTHORS

Sarah Reid

Lead author Sarah Reid is an award-winning Australian freelance travel writer and sustainable travel specialist with a passion for adventure and immersing in nature. She has spent more than a year of her life sleeping in a tent everywhere from Africa to Patagonia, with her favourite sleeping-under-the-stars adventures including a four-month camping odyssey along coastal Queensland, and glamping just steps from Western Australia's Ningaloo Reef. Sarah writes for top travel titles in Australia, the UK and North America, and has also written or contributed to multiple travel guides and reference books including The Sustainable Travel Handbook, published by Lonely Planet. Read her work at sarahreid.com.au.

Tasmin Waby

Tasmin is a London-born writer with Kiwi *whānau* who grew up in Australia. She's had the good fortune to drive, hike, cycle and canoe (with her camping gear in tow) to some of the best wild places in Australia and New Zealand. More recently she's been pitching up at music festivals in England and wild camping in the Scottish Highlands. Tasmin loves nothing more than falling asleep in a one-man tent listening to a river. Since reading Walden, she's harboured the living-off-grid-cabin-in-a-forest dream (although perhaps with enough solar power to run a top-quality coffee machine).

Andrew Bain

Andrew Bain is a Tasmania-based writer specialising in the outdoors and adventure. He once set out to see how many consecutive nights he could spend in a tent and was foiled at 120 nights by gale-force winds at Nordkapp, Europe's northernmost point. He's cycled around Australia, a 14-month journey spent mostly camped by the road and is a regular wanderer of the wilds with a pack and tent on his back.

Andrew is the author of Lonely Planet's A Year of Adventures and Hiking and Tramping in New Zealand, and also the author of Ultimate Adventures Australia. His favourite five-stars are the ones he can see from his sleeping mat.

Under The Stars Camping Australia & New Zealand
April 2024
Published by Lonely Planet Global Limited
CRN 554153
www.lonelyplanet.com
1 2 3 4 5 6 7 8 9 10
Printed in China
ISBN 978 18375 8173 3
© Lonely Planet 2024
© photographers as indicated 2024

Written by Sarah Reid (QLD, NSW & ACT, NT, WA,
SA), Andrew Bain (TAS, NZ South Island) &
Tasmin Waby (VIC, NZ North Island)

Publishing Director Piers Pickard

Gift & Illustrated Publisher Becca Hunt

Senior Editor Robin Barton

Editors Anne Mason & Vicky Smith

Art Direction: Daniel Di Paolo & Emily Dubin

Layout Designer Lauren Egan

Indexing: Vicky Smith

Print Production: Nigel Longuet

Cover Illustration by Ignasi Font

Lonely Planet Global Limited

Digital Depot, Roe Lane (off Thomas St),
Digital Hub, Dublin 8, D08 TCV4, IRELAND

STAY IN TOUCH
lonelyplanet.com/contact

Paper in this book is certified against the Forest
Stewardship Council™ standards. FSC™ promotes
environmentally responsible, socially beneficial and
economically viable management of the world's
forests.